ORGANUN

FAREK J SADON

authorHOUSE

AuthorHouse™ UK
1663 Liberty Drive
Bloomington, IN 47403 USA
www.authorhouse.co.uk
Phone: 0800.197.4150

© 2017 Farek J Sadon. All rights reserved.

No part of this book may be reproduced, stored in a retrieval system, or transmitted by any means without the written permission of the author.

Published by AuthorHouse 08/29/2017

ISBN: 978-1-5462-8050-7 (sc)
ISBN: 978-1-5462-8081-1 (e)

Print information available on the last page.

Any people depicted in stock imagery provided by Thinkstock are models, and such images are being used for illustrative purposes only. Certain stock imagery © Thinkstock.

This book is printed on acid-free paper.

Because of the dynamic nature of the Internet, any web addresses or links contained in this book may have changed since publication and may no longer be valid. The views expressed in this work are solely those of the author and do not necessarily reflect the views of the publisher, and the publisher hereby disclaims any responsibility for them.

GUIDE TO THE READERS OF THE TEXT BOOK

Please find the Text, Syllogistic and Scanned Illustrations, Elaboration and further simplified details that assist the reader; personally imagine how Organun Proposal Works and how individuals and Industry can produce Such an Instrument to obtain Vision and Instantiations, simultaneous transmission, Translation & Interpretation.
And how this Method involved and applied in the process of Meditation and Self skill Extraction Obtaining Qualifications via tunings to the text and following the progression **in contrast to their individual minds Via the Operation of the Axioms and Common Notions; their own apprehensions** and their own culture backgrounds which is the foundation of their own philosophy; their own arithmetic logical

The way is naturally their minds tuned to their own Stars

Plus how individuals and groups obtain capability to express their own vision and how they are becoming in the possession of their own methods of contemplation in to inferential and deferential realities thus capability to produce out of their own arts naturally planted in them out of the abundant in nature. Man and the elements are contained in an Environment, earth, air, water, fire, light Plus the Faculty of Prediction which is planted in Human Heart as a desire and ambition which is is hold in the future.
Every Individual being contains a value and the Algebra; The Obligation of Life; the Breathing system in which the organic being is extant the faculty of Human mind pressurises human Physiological system to extract out

of his own value because it is the only way that can fulfils Individualistic Ambitions and that is in itself the Cause of human probe and human suffering.

You are Ether a Slave Labourer or an Affluent and Liberated
Having not your own vision
You have the first option

THE INTRODUCTION ON CONTEXT OF THE TITLE

BRIEF HISTORY OF THE TITLE ORGANUN

Organun is a Title given to the Instrument of
 a. Human Mind
 b. The logic of human mind
 c. The instrument of time
 d. Human Brain, Senses and psyches

The title terminated and coined by Ancient Commentators, "organun is that container in which Aristotelian logic is contained"

Brief History of Aristotelian Logical Works Titled under Organun: The Instrument of Philosophy:

Aristotelian Logical Works are equal to Exegeses
Exegeses is a process which is **in equal ratio with to read from** a Text Book, record, Holy Geometrics, or Holy Scripture and that Text Book is called Book of Vision (Kitab Al Mubeen, Islamic translation)

 - **Metaphysics**

Phenomenon; *such kind of Vision*, (Book of Vision) (Kitab Al Mubeen, Islamic translation)
Has been Exposed to various Interpretations and expressed with several forms of translational means in the course of progression of human thought such as:
Metaphysic, Dream, Essene, Episteme, knowledge, Epistemology and inspiration thus Aristotelians are Human Diasporas that in

Virtue of Possessing Vision they have obtained Vision via the Element of Human Mind which is a common notion, the element that is in Virtue of possessing it we all have IT. thus Diasporas are qualified literates who possess the faculty of Vision (the ability to read from the Universal horizon) and that separates Aristotelian logic [in the process of revision of platonic philosophy] from mystical, sophistic and homonymous Prophesy which is enveloped in religion order thus Aristotelian logic is always contained in *Organun the instrument of philosophy* <u>The Element of Human Mind the substantial qualification</u>

We Have Scientific Knowledge When We Know the Reason the cause why the thing is? The answer because of this and cannot be otherwise such propositions are coined as non-contradictories in the content of speech thus the object of scientific knowledge:
   a. Only the necessary the case can be known scientifically
   b. Scientific knowledge is equal to the knowledge of causes

Thus to start with observation process that at any rate one formula of science consists in the possession of demonstration and that is in itself a **scientific deduction** thus the scientific episteme Entity is in Virtue of Possessing It We all Have episteme and Knowledge; and that includes every single individual mankind on our planet Earth no matter race or land because racism and patriotism are formal objects of the Essenes of mankind and it is the Essene in which we are all equal *Black or White*

The text book concerned with basically two tasks:

1. To pronounce the origin organ of Demonstration and the demonstrative means
2. Answering its possibility and challenge of application.
   Demonstration is equal to a Deduction in which the Premises are: True, Prototype or Primary and Immediate without Middle and that Premises is Equal to time and time equals to human apprehension See Greek Names, AMESA Medusa with having Hydra Shapes or pentagonal Star.

- Religion Organ

In Religion organ the phenomenon is described by Inspiration with reference to Metaphysics, heavenly mystical shrines or ritualistic holy places in the domain of demonstration and cultural tradition Thus methods introduced to obtain vision or to reach that Horizon via which one can imagine the holy record and to learn how to read from it and how to manifest the holy scripture are methods such as *to pray via ritualistic* Practices and demonstrations these methods includes oriental Buda-Indo kind of preaching and tribunal scholastic oriental mathematical Illustrative means known as Mandalas

- Theologies

Is an academic approach toward metaphysics, starts from theory of creation then enters into logical reality and the realm of theoretical physics And it is the theory of creation that to which Aristotelian Metaphysics is referred, the Dark Part of Greek Philosophy and limitation of Aristotelian logic. In Greek philosophy Dodecahedron Identity is not defined or objectified nor physically constructed and that is the limitation of Aristotelian logic. The Dodecahedron Identity is enveloped in mysticism in Aristotelian logic; as it acts to ornament the universe. See Platonic philosophy, and the dilemma of Postulate Five in book of elements.

Brief History of Holy Geometrics, Global Vision or The holy text

The programming language enshrined in the book of Elements is to logically construct the five regular solid state physics

    A. Fire
    B. Water
    C. Earth, Cubic Gravity
    D. Air
    E. Light [Systems: Observation + Communication]

Production of light is in equal ratio with to reach enlightenment and to reach that is to obtain vision and obtaining vision is equal to human ambition (seeds planted in Human Hearts are desires and human ambitions is their fulfilments which are hold in the future thus future predictions are nowhere apart from the progression of human evolution towards an End which is primarily determined and inscribed in time.

Time equals to a permanent cycle that contains 3Dimentional objective reality: the tension Past, Now and Future which are defined by time and the physical reality of time is equal to its faculties the power of imagination and the faculty of vision sight, seen, sound and hearing and the tensions, human inner feelings in the form of pulses and Seen –Hearing in the form of beats The Organun text book is to introduce and to objectify the identity of the holy (record-geometrics and the method of exegeses to read from the record and to deduce theorems, to construct, express and create The term Algebra used since the book of elements translated into Arabic within the emerge of Islamic rule the term indicates human faculty of imagination which is in equal ratio with faculty of thinking (Mathematics-analytics the process of thinking)

> to think to imagine, to read to see to deduce thus algebra (Power of imagination) is an Entity that is in Virtue of Possessing It We all have the ability to think to imagine the faculty of imagination is concerned with elements and with operations, prior to their definitions they obey certain laws called postulates any means of interpretation may later be given to these elements and operations will if they satisfy the postulates also satisfy the theory of Imagination. Every universal law of thought, Religious, Legendary, Mystics and all forms of speech, tails and narratives are constructed via mathematical means, which is the ultimate power and it is the subject and predicate for itself the element of Human Mind.

Text book concern
Text Book concerned with:

> Optimal Solution of Inferential and Differential Equations and that entity is Organun the Title of our study, the Text Book

Logos:

All forms of speech called logos and are constructed via mathematical means, which is the ultimate power and it is the subject and predicate for itself the element of Human Mind.

Human Ambition equals to reach the ultimate knowledge to knows the reason and the power that directs the proposition of **to be or not to be** by Necessity and it is like that necessarily and Science is the only mean that is able to determine what is hold in the future and future occurring events. Physic is the only Science and platonic curriculum is the only philosophy Thus philosophy = Love to learn to know and that is in itself equals to the realisation of the law of Identity and self-recognition then emerges the proposition of Almighty God Is Living Man and I am the superior thus free.

Faculty of Vision
Faculty of vision is defined as the ability to:

    a. Think
    b. See
    c. Hear
    d. Feel
    e. Utter
    f. to Xpress or to Demonstrate

Above Mentioned Faculties:
- Are extant only in Organic Structure or Composition which are Combined together in one Single Autonomous **FORM or Body**: that has a Mass, Mass Number and Volume and Equals to [**Equation**] **Math Symbol for the equation encompassing this is (=)**
- **A Living Organ**
Both **animal and a man** are **physical realities** born of **Equations**. Living organs are **Optimal Solutions Inscribed in a System of Inferential and differential Equations** and that is what **the concept of a cycle** attempts to **demonstrate** in addition, the **concept-form** of a cycle Terminates: via 3 facts, {Centre, Sphere

and Straight line or lines}: are equal to texts and Vocal Expressions and are merely concepts (exhibiting axioms that combines centres into spheres and exist in a *permanent, continues* and Mutual Centre Sphere Relation).

Categories:

A Category is Equal to a Text from the Organun that enumerates Human apprehension,
Human apprehension equal to a subject, or a predicate for any proposition in The permanent cycle of time

Enumeration

Categories, the actual Numeric Basis of science, the faculty of **enumeration** the ability of human mind to enumerate;
   a. to express
   b. to know the number of his own organs
   c. To count
   d. To commensurate
   e. Finally knows arithmetic logic (Elm Al- HISAB Islamic translation) to account for its own Nature,
   f. All universal laws of thought, Religion, Legends, Mystics, all forms of speech. Are constructed via mathematical means, which is the ultimate power and it is both subject to and predicate for itself the element of the Human Mind.
   g. **The Text equals to the Faculty of Enumeration and that faculty is of itself equals to the power (the Optimal Ultimate, Power and that power Organ or Element is not in the Space But it is in the Human Head**

Categories 1,2,3,4
   a. One dimensional
   b. Two dimensional
   c. Three dimensional
   d. Four Dimensional continuum (Time)

Coin Shape equals to (2sides and a line in between) and equals to 3 Dimensional objective reality

And number 4 equates to time ant that of itself equates to Human apprehension

Prior Analytics:

Prior to the intellect (the Thinker or the thinking entity and its process) exists a System of Breathing, inside of which living organs are extant and that in itself; Synonymous with, The Essene or Soul….. Mathematic Symbol of the Essene is equal to Zero, **0.** **The conclusion is that; prior to** the Thinking Element **that contained in the equation Exists [System of Communication and Observation] the universal environment, the cosmos that container that Thinker the Analyser, the Bio …are Contained.**

e. Time
f. Space
g. Motion
h. Then exists: the _**Thinker**_ the thinking faculty, the Thinking Element

Post Analytics

Equal to the Element of the Human **Consciousness** Placed on its **Sub- Consciousness**, the prior analytics, and that is in itself equals to **[Mathematic Probe]** thus it is an enquiry via Prior analytics into Future Predictions.

The Text Developed by Organun Ltd is equal to organic structure and composition of Human Mind thus equals to Human Logic and Equals to Book of Vision KITAB AL MUBEEN Islamic translation

Organun Proposition

A. Everything that Exist equals to structure <u>or</u> composition in time Human apprehension Equals to time in Structure <u>and</u> Composition Apprehension

B. Everything that exist equals to pendulum *or* Clocks in time Magnet and Magnetic fields is Equal to time in Pendula *and* Clock apprehension

C. Everything that exist = points or lines in a pentagon
Human Grip or apprehension = to time in points and line apprehension

On interpretation

> It is the Human inner feelings that are on interpretation and that equals to the progression of human cognitive power from particularity to universalisation or generalisation and it is called syllogistic

> Orientations and Law of Identity
> North, South Polls of the magnet, and Head to Feet are in equal ratio with Sky, Earth, and States: (Gravity and levity) gravity force toward the earth and levity force toward the sky example: Smock, Gases and Time Space Motion Relation in which the Living Man, Human apprehension. Situated INSIDE the Transformation operation of elements, the four Dimensional continuum elements see Entropy & the Second Law of thermodynamics
> And it is that the concept of Cubic shape postulates or trying to demonstrate and it is a text of the Planet Earth and equals to cubic equation or logical construction plus the substitute number 7 and that make the Essene of heptagonal in the crystal system

Topics
Related and tangible momentums and they are time topics and are ever equal to present **Position and Definition of Faculty of Vision**
Human mind itself is a kind of thing it is called Energy converted inside the transformation process for Example, into sound and hearing media, sight and seen media....
The human Element itself is the constructor of our senses thus energy equals to the physical definition of the Expression [Essence] which is enveloped in mysticism in the context of Greek philosophy and Energy Equals to Mass Times Square of Light and Equals to the Faculty of Vision

a. Man and animals are both breath takers
b. Thus the one who can think and see and say is Situated in the Cycle of Elements And the Elements are:
F. Fire
G. Water
H. Earth, Cubic Gravity
I. Air
J. **And Light** [Systems: Observation + Communication]

Definition Equal to Point *Which has No Part* **and Equals to** *One Dimensional Being* **and equals to** *Position* **and position Equals To** *Time Place Intersection*

On Sophistic Refutation
To take side with that is in between, the golden mean, the episteme to avoid the state of contradiction or Identity but to hold with the state of nun-contra identity, the state of perpetual peace, not to wrong Measurements or tools in the process of **critique of pure judgment** or Construction of speech

This requires choosing right tool to obtain a right path the instrument of philosophy Organun

Poetics:
Human Feelings Natural Feelings Such as Passion and Pain and are equal to numeric square roots inscribed in forms called **quadrature, quadric is a quarter of a square** and equals to time space intersects, are called **clocks and spins** thus quadrature equals to a Pyramid and Pyramid is equal to (**structure or composition in time**) (**Pyramid apprehension equals to Time in Clocks and spin** *Grip, Understanding or apprehension*) Structures are [Weights-Grammar entities] defined as Tension, momentum, gravitation… and they are equal to indexes of Human Feelings. A **Pyramid** equals to Optimal Equation and equals to the Cubic Equilibrium. Squares are Prototype Forms of Human Feeling and are equal to four Numbers [4X4 = 16 – 4 = 12/4 = to a polyhedron (geometric forms) with four Equal faces, each of the faces is equal to the expression of the other 3 Numbers by one of the Numbers. the four numbers substituted in Arithmetic-Geometric Operations construct a compass in the form of **tetra hedron** See the Text Book

Definition of the Organun Text:
Element text Equal Biometric
Regularity and regular forms

Pentagon is the form that contains the Elements Tetrahedron, icosahedron, Cubic, Octahedron and Dodecahedron thus Pentagon is a form that every other true forms of Existence Emerges through The process or progression is the Program of elements and the procession of Human Cognitive power from their Particularity to their Universalities or Generalisation And that will be the method that we are going to present in the text book then the readers will be provided with a guiding tool to Understand Topics that Exist and then themselves be able to extract qualifications that identify how far those topics are related to them and how Physic, the natural mechanical element is in mutual relation with human apprehension.

Rhetoric

True Propositions, Scientific Statements, optimal resolution in demonstration, utterance or expression True forms of Speech and Declaration the Gospel

[Al Hadith in Islamic translation See Ibn Khaldun Khaldunian Islamic scientific History the Introduction al Muqadma]

A persuasive speech that to receive an optimal agreement also with having the power and basis to hold tied to its straight path that not to be deflected by the attraction of sceptical argumentations And it is a power to which its base is raised in order to become a proposition or assertion. Thus rhetoric is equals to an assertion that built on organic logical structure and composition.

Guiding Introductions
a. Categories: equal to the Text:

The text equals to the faculty of Enumeration

And that faculty in itself equal to the power (the Optimal Ultimate, Power and that power Organ or Element is not in the Space But it is in the Human Head Example Electric Charge is Not in the Iron or in the Piece of Magnet via which induction Method is applied.

The power is in fact contains in our head and that Magnet or the metal which becomes a conductor and conducts the current and the electric charge are no more than a reflectors reflecting what is contained in the human mind and Head and that is Human Head Which is inscribed or incubated in the cycle the concept (defined by central Point which is situated in a centre of sphere or has a boundary i.e. the concept of the centre and sphere

Thus anything spherical or from sphere are no more than [Mirror Reflecting processes]

And the rest of the Power Contained in the Head: the Ultimate power. The category of 1234 (four Numbers can be imagined as right, left and the line In-between (the Separator and the conductor of the two Sides) that forms A Coin Shape: the Physical reality of a Piece of Magnet and the physical realty of every Solid. Thus Solid is equal to a plain Figure (real nature of a piece of Magnet) which has been used as a tool throughout human history in Navigation, Aviation, electricity production and finally the entire digital industrial technologies in modern times.

Let represent those Numbers By A, B, C and D

A, B + C are equal to 3- dimensional phenomenon and D is Equal to Human Inference

That is why Vision and Electricity has become the core context in this universal study or

In the context of this Text Book

    b. On The Direction of the Current:

> *the current produced via induction and the light that emits via Force times Resistance process, the light produced inside tube or a law pressure Air container, is Proportional to the magnet, the conductor metal and the human head*

And that is eye, light and magnet relation, and this is in equal Ratio with Magnetic power of the Earth and Heavenly bodies or boundary attractions, the conclusion is:

Head to feet Ratio Equals to Ski and Earth Ratio the entity in between No one but the living organs in which Man is the Optimal Living Organ and the four dimensional Continuum equals to the cycle or plain circle *described in book of elements.*

Gravity, Levity, plus the four dimensional continuum are Equal to Six Dimensions
And the number seven is the ultimate position that Knows Every Thing. Thus
1234 or a. b. c. d are equal to the 3 dimensional organs plus time and time is equal to human apprehension.

On the mean that this text book provides:
Pentagon Described in the topics of this book See Text Book.
Equal to that Text [a five angle stare, Pentagon with equal sides and edges]
Pentagon Is Text Confirmed by Human mind, the Faculty of Time.
By the term (In Time) which frequently used in this text book We Mean in The eye, Opinion or view of Time.

Propositions:
Everything that Exist equals to structure <u>or</u> composition in time
Human apprehension Equals to time in Structure <u>and</u> Composition Apprehension

Everything that exist equals to pendulum *or* Clocks in time
Magnet and Magnetic fields is Equal to time in Pendula *and* Clock apprehension

Everything that exist = points or lines in a pentagon
Human Grip or apprehension = to time in points and line apprehension
Thus the text equals to Biometric Text
The symbols used in the pentagon text are Numbers Odd, Even ; lines or points, Points or Angles See illustrations:
Thus the text equals to the dictionary and equals to time and that is the Special Basis of human mind and equals to Human apprehension and are equals to Definitions, Postulates Axioms and Common notions. described in Book of Elements

In vision of this text book

Organun

- Algebraic Power is a Mean, Median, Optimal Power and equals to the element and that element is human mind itself [Brian, Senses and psyches] in the cycle of time

The power of Imagination is the Element that converted inside the transformation process and it is recognised and called time with reference to Human Mind (Human Mind is the referendum and context that contains the faculty of Confirmation and Denial)

Human mind itself is a kind of thing it is called Energy converted inside the transformation process for Example, into sound and hearing media, sight and seen media....

The human Element itself is the constructor of our senses.

In the Essene our senses are Mirrors and they are in mirror imaging process.

Our senses are demonstratives and demonstrate our inner feelings

See illustrations Regarding the heart and heart beats

Music is hearing Notes Detected by Human faculty of Hearing and it is called inspiration in Gospel or legendary forms of speech the instrument, the element that converted into instrumental musical organs are called Gabriel, or temple of inspiration in religious and legendary forms of speech and Melody is the Greek name for the organ musical instrument and it is the inspirer himself, the music maker and the intellect value that has been converted inside the system of organic transformation.

Thus to reach inspiration or to receive inspiration is to tune mind to the melody.

Today, in the advent of the current system of digital construction process and industry it is possible for that to be achieved and that is not a mystery, our five senses are mirrors and are constructing five arithmetic compasses and five top Geometric compasses. See The Text Book

Pyramid Shape Equals to a constant ratio in the system of observation, breathing and communication.

The Pyramid its mass number equals to five $5 \times 5 = 25 - 5 = 20$ thus 5 equals to the five senses substituted in commensuration or Counting Process and are equal to the mass Number and 20 is equal to its volume number that constructing an icosahedron with 20 triangle basis

20/5 = polygonal/ Pentagon with 5X4= five Squares each square contains a sense as a point and what contained in the squares is equal to that particular Sense Expression in the square of the other four mirrors.

Pentagon is equal to the text that all other 5 regular polygons and polyhedrons are emerged

Tetrahedron, icosahedron, cubic, octahedron and dodecahedron
There is no more regular polygon or polyhedron

The programing language enshrined in the book of Elements is to logically construct the five regular platonic solid state physics and they are forms of thought and called Concentration of mater and its Identity equals to thought Forms (forms of Thought and are physical Identity of Grammar- or Weight, to concentrate is to contemplate is to think seriously, to concentrates analytically. Thus regular forms are forms of concentration having the same ratio with concentration of water into ice or Air into octal, fir into tetra and earth into cubic solid with six faces and dodecahedron with 12 Hydra or Containers And are all make the essence of the crystal system.

All living organs that can analyse are contained in the systems of:
   a. Breathing
   b. The observation and communication system that contains the observer and observed, speakers and listeners and their mutual relationship in cycle of time.
   c. Air median Air- dynamic
      And it is a time and place, where the Identity and the entity of the thinking Element is positioned and both phenomena of Vision and Electricity are exist
      Squares are scales of mirror forms and are equal to containers of our senses
      And squares are mirrors that contain images of our senses with ratios of 1/4
      And that equals to a pyramid a constant shape which stays invariant permanently in the cycle of Breath, Observation and has equal edges and lines or sides:

5 equals to its weight Number and 20 equals to it volume No. thus its form equals to a shell composed with 20 numbers. In the case of our sense compasses, are five regular standing clocks we have five pyramids each times 20 are equal to one Hundred (100) and that constructs the time table of hundred which has been not recognised or identified by any mathematician before the author of this book.

- **Arithmetic Logic Means Logic of Earth**

Language of Earth and that Means What the Earth Expresses
  a. **Rock**
  b. **Water**
  c. **Sand**
  d. **Zio**
  e. **Bio**
  f. **Animal**
  g. **Man**
  h. **And what is in the Future or hold in the future: Prediction**
  i. **Predicament/** See the Categories Listed in Aristotelian Logic

And are all contain in the cycle of time

History equal to time repeating itself

Geometrics Means what Expressed by Earth or what is on the top of the Glob of Earth

In contrast with Human Mind

Conclusion is equal to 3 Numbers

A, B and C

A= to Expresser, Expression and operation relation in time, the element of recognition the entity that recognises and knows and inside the operation transformation of express and expression process, The Faculty of Vision.

The tools that we are intending to construct or to be constructed are Arithmetic compass which is equal to Expression Compass, and Geometric Compass, the Expressed Compass. Top compass: Geometric Compass and the Sub Compass the arithmetic Compass and are both Arithmetic and Geometric Means.

See, the text book, Illustration figures. On Cubic, incubation, Octahedron and dodecahedron

Illustrations on the Method Used in this Study:

1. On the concept of Organun, the title of this **Text Book**.
2. Faculty of vision:

Ability:
a. to Think
b. to See
c. to Hear
d. to Feel
e. to Utter
f. to Xpress or to Demonstrate
g. **Above Mentioned Faculties:**
h. Are Exist **and only Exist** in Organic Structure or Composition which are Combined together in one Single Autonomous **FORM or Body**: that has a Mass, Mass Number and Volume and Equals to [**Equation**] **Math Symbol for the equation is (=)**
i. A Living Organ: **an animal or a man,** are **physical realities of Equations**. Living organs are **Optimal Solutions Inscribed in a System of Inferential and differential Equations** and that is what **the concept of a cycle** trying to **demonstrate** Also the **concept-form** of a cycle can be Terminates: via 3 facts, {Centre, Sphere and Straight line or lines}: are equal to texts and Vocal Expressions and are merely concepts (exhibiting axioms that combines centre to sphere and are in a *permanent, continues* and Mutual Centre Sphere Relation).
j. Therefore **any Mental Act or Thinking Process Including Reasoning** exist and **only Exist** in the domain of **Breath taking Species**, **Living Organs** thus prime to thinking Element and its process Exists a System of Breathing, inside of which living organs do Exist, and that in itself; Synonymous with, The Essene, Soul, Breath….. Mathematic Symbol of the Essene is equal to Zero, **0. The conclusion is that; prior to** the Thinking Element **that contained in the equation Exists [System of Communication and Observation] the universal environment, the cosmos is that container that Thinker the Analyser, the Bio …are Contained.**

a. **Time**
b. **Space**
c. **Motion**
d. **Then exists: the _Thinker_ the thinking faculty, the thinking entity**

What Separates Animal living organs from Man Living Organs is that, Animal can feel but do not recognise itself, a Pony Do Not Know What Pony Is.
Man Recognise what is Man and Manhood
Animals can utter Sounds.
But Can Not Utter Words
Man Can Utter Words Thus Able to talk
Man Able to Feel and to Imagine then to recognise himself. However the Question is: Has Man, so far arrived that conclusion, reached to that realisation and has he really recognised himself. The answer is No and that is what which is hold in the future when man managed to realise himself there will be no more ignorance and that stage is: State of Nun contradiction Since this recognition in itself is Equals to the Method of solution of questions in the theory of Probability, see Principia Mathematica, Newtonian Propositions, and calculus
The state of perpetual peace and Optimal Resolution, to have **Vision is in equal ratio with to have literacy to have no vision is in equal ratio with to be illiterate or Blind** Exclamation mark is that, are we illiterate? In the current age the answer is yes simply because currently an international curriculum of Science and Education is not available! What is now available as a Frame or curriculum of education system stands or Based on a principle called Scepticism and the state of scientific judgment is in the state of uncertainty; Science is the one who knows the reason and the power that directs the proposition of to **be so** By Necessity it is so by necessity or necessarily and Science is the only Element that is able to account for future and future occurring events. Physic is the only Science and Platonic Curriculum is the only philosophy
The current global systems of teaching unable to recognise, Special Numeric Basis of science thus provide more than one reason, for the occurring events in its Critique of Pure Judgment, assertions and justifications, there is no universal law and Scientific Method for an optimal solution for the Questions in the theory of probability.

Categories, the actual Numeric Basis of science, the faculty of **enumeration** the ability of human mind to enumerate;
  h. to express
  i. to know the number of his own organs
  j. To count
  k. To commensurate
  l. Finally knows arithmetic logic (Al- HISAB Islamic translation) to account his own Nature,
  m. All universal laws of thought and all forms of speech. Are constructed via mathematical means and by the ultimate mathematician the element of Human Mind which is the subject and predicate for itself in time

The Topic Sceptical Defined

Inferential Equations are inner fillings having, the ability to imagine their own inner (feel- states)
For example, Passion to feel high or low….the Inners faculties, possess power to imagine such states via construction of forms or texts for those, Inner States and Tangible Moments.
And that is what we **mean** by the **Power of Imagination**. Thus the universal so Called Holy geometry Points, Lines, Shapes, surfaces, and 3 dimensional forms are all texts constructed by the power of Imagination. Thus functions Such as to think is in equal ratio with to read from a record things like Eye, sight and seen the 3 are taking place in a universal operation of time.

   3. **Position and Definition of Faculty of Vision**
      c. Man and animals are both breath takers
      d. Thus the one who can think and see and say is Situated in the Cycle of Elements And the Elements are:
         K. Fire
         L. Water
         M. Earth, Cubic Gravity
         N. Air
         O. And Light [Systems: Observation + Communication]

4. **Definition Equal To Point** *Which has No Part* **and Equals to** *One Dimensional Being* **and equals to** *Position* **and position Equals To** *Time Place Intersection*

The Context of time equals to a Cycle in which Human Head is Incubated, and this, Synonymous with functions terminated by **Inscription, Entangled, Positioned or Situated.**

   a. Inferences are organic Axioms that are to differentiate two sides of one coin and to combine both sides together in a form or Shape of a coin simultaneously.
   b. To Burn is in equal ratio with to feel to separate black from white, Example to burn to release Ashes, grey equals to the black-white composition to separate black from white and are called the deferential equations
   c. Thus prime to fire, the element with ability of *to burn, the element of fire* the actual Identity of the thinking Entity, some other element or Energy being spent or being converted into transformation and it is inside the process of transformation and it is equal to (prime time took place before) See (the dilemma, Entropy and second law of thermodynamics.
   d. The 3 Dimensional Forms: are 3 differential axioms.
      - The one which separates the two sides of a coin is the one in between Separator and combinatory it is the entity that burns thus the entity that feels

   The category: 1234 (a, b, c, d) a or b are two states like day and night then c is the one who differentiate the other two existing states thus it is the one who separates and that equals to **Strength** that combines Both States. Total 3 are constructing a unique form of **Coin** two Sides and the line Combiner in between. The third line the Line In Between is Equal to an Arrow Directed at, Ends up in the centre of the cycle and that Particular position, the Centre, is equal to the position of human head Human Head.

   [And that defines North, South Polls of the magnet, and Head to Feet are in equal ratio with Sky, Earth, and States: Gravity, levity gravity force toward the earth and levity force toward the sky example Smock, Gases and Time Space Motion Relation in which the Living Man, Human apprehension is Situated INSIDE a Transformation operation of elements: see Entropy & the Second Law of thermodynamics, Boolean algebra and Empirical Views, Anglo -German Mathematical Probes in the process of construction of Arithmetic and Geometric Logical Means, *Compass and straight line*

imagine two ends of one Straight Line, for instance Sounds that produced in the sphere, space are recognised by human Hearing Media and the line that Connects sphere with centre point i.e. The position of Human head and the distance in between, Head to the sound event thus hearing-seen axioms themselves are Axioms or Organic Structure entities: now we can imagine sound postulates and the postulate itself is a sound constructed via axioms and are common notions for the axioms: are the organ entities, Magnitudes are proves of sound and are organs or hearing axioms; Music is sound or vocal expressions in other words music equal to sound on interpretation. Thus our hearing-seen mediums are means of interpretation and Understanding simultaneously and things, in space, Outside Human Head are postulates and the axiom or common notions are means of recognition Example Sound and Ear ears are the entity that recognises the sound and it is itself equal to mean of interpretation Thus c Times d, (is the point of seen: Eye and Light Ears and Sound). Intersections

The physical reality of a Coin which is physically exists in the realm of Nature or Demonstration; exactly equal to a piece of magnet two Sides and the side <u>in between</u>

5. **Brief History of Book of Elements:**
    a. Euclidian and nun Euclidian Geometry Dilemma
    b. Postulate Five Dilemma
    c. The 3/2 Ratio

All parallelograms (illustrated in the Illustration Figures) are equal Parallel lines and are Equal Sides of a Magnet and the third arrow line that directed at human hearing media: the head, Back Head and front the Face. **This is an intersection relation, with the current or Electricity produced via induction, Tools used in the process; is [piece of Magnet and current conducting metal] then production of Light, radio activity,** radio waves, TV screens, CRT Cathode Rays and the relation, Electricity and its interaction with human Faculty of vision, the light produced in a vacuum tube is in common notion, relation, interaction with the piece of magnet and our sense of vision (Light, from the sun or air empty tube) (Piece of magnet via which the electric current, electric charge produced) And human Eyes or Hearing Media the light produced via magnetic media is proportional to human Eye Ears and the piece of magnet and that ratio equal to **3/2**

This text book introduces the physical reality of the three Dimensional Forms via below mentioned physical Means:

a. Electric Current Produced Via Induction
b. The Nature of Conducting Metals
c. The Law of propagation of Light in Empty Space
d. The Light Produced Via the process of Current Times Resistance (Heater and heated Filament, production of direct current then production of light, the Entity with Dual Nature, Wave particle Dilemma) laser beams, radio waves, microwaves the total application in the domain of communication and observation

Technology thus we have three things in the hand as a **Tool of Science**, Magnet, Metallic Conductor, Empty Tube, Optics, Computer TV Screen and a magnet in the hand that can deflect the light produced inside an empty Tube or container and the third line that connecting human mind to the event taking place in the space, outside of Human Body or to the spot of light produced in the CRT or Computer Screen.

Inference Defined.

6. **Element text Equal Biometric Review From The Past Times: the Antiquity**
   Power of Imagination
   Physical Reality of Book of Elements
   Essence or the Intelligentsia
   Elements or TEXT of the Intelligentsia

On Platonic Solid State Physics/ the Logic of Antiquity
- Tetra Hedron Text, Form, Formula and the Law of Fire
- Icosahedron Text, Form, Formula and the Law of Water
- Cubic Text, Form, Formula and the Law of Earth
- Octahedron Text, Form, Formula and the Law of Air
- Dodecahedron Text, Form, Formula and the Law of Light
- Vision: Human Eyes Relation With Light
- Human Hearing and seen Media relation with the space
- The cosmological constant Equals to a container Called Cycle
- the cycle is composed with only two things Centre PLUS Sphere and the mutual centre sphere Relation in time
- A Diameter of cycle (A diameter of a circle is a straight line through the centre and terminating in both directions on the circumference which bisects the cycle and straight lines from the centre to the circumference is called radius; plural, radii.
- **A Diameter Equals to Text or Biometrics of Human Feeling and Equals to units of our Feeling or Consciousness and are Viewed or imagined by Seconds In the vision of time and in fact they are our feelings inside circular transformation and are equals to the strength thus the strength of gravity is our feeling and are the labour that every living creature must perform; thus it is in itself algebra or Syllogistic that progresses from an existing prime provision to the determined end and by necessity an obvious example of this is the obligation of life we all love to live and born without our choice.**

Organun

- o Pendulum Motion Equal to Its Physical Demonstration (the Strength)
- o Pendulum Motions could be imagined as Bell having a handle in-between two only two polls
- o The Physical Reality of Pendulum Motion is Equal to a piece of Magnet
- o Pendulum motions are represented by lines in Book of Elements and a straight line is an equal distance between two points
- o Thus piece of magnet which is equal to two polls and magnetic field inside the one called Magnetic Field, Radio Active Median, electrodynamics, and mechanics are all terms describing the Magnetic field our Body System Earth and Sky and the four dimensional continuum: ***Centre and Sphere and Centre Sphere relation***

- o **Diameter and strength of gravitation identified**

7. **On Postulates: Root of I/2 Ratio, physical reality of the point and the Function of x**
    a. To draw a **straight line** from any point to any point: To have a piece of Magnet
    b. To extend a straight line for as far as we please in a straight line: to cut a piece of Magnet into infinite pieces yet equals to a coin and that equals to ½ ratio
    c. To draw a circle whose centre is the extremity of any straight line, and whose radius is the straight line itself.
    All right angles are equal to one another: Right Angle is Human Hearing Media times, At Horizon, Example, to look through Pair of Glasses and that is equal to Human Semi Sphere

Back and the front Human Logic: Face Back Head and Face Front Watching, Looking, listening, talking Direction This Ratio Is Always Equal to Pyramid Shape

Five Equal Right Angles and are equal to 5*5=25-5= polyhedron with five Regular Squares Pentagonal each of the squares contains one of

the Right Angles thus a right angle that contained in a square is equal to expression of the angle and that shape is equal to X all X shapes are forms of multiplication and are equal to a common notions and the axioms are no more than the two numbers multiplied by each other
If right angle is light then what contains in the square is equal to reflection if it is sound or voice then what is contained in a square equals to eco and received in both cases as hearing or seen simultaneously each square contains a tetra for tetrahedron
$4*5$ squares = to 20 tetrahedron or $20*3 = 60$ (20 Hedra with 20 Triangle faces
or basis or seats thus in any triangle $3*3=9-3=6$ thus every triangle has six faces 20 triangle$*6= 120$ in five squares of a pentagon at equal rate 24 each square and each 24 contained in four sections **Quadrature** 6 each: a star shape polyhedral a pentagon equals to five $4*6$ polyhedron and Equal to five four hexagonal Polygon **and that equals to the plain of human head.**

Postulate five: a straight line that meets two straight lines makes the interior angles on the same side less than two right angles, and then those two straight lines, if extended, will meet on that same side.

Postulate five defined that means the dilemma of <u>nun Euclidian geometry points Function of x are illustrated</u>

8. Dodecahedron: the Fifth Element

Syllogisms: Light Entities (Particles) or Visible Light
Visible Light the Dodecahedron Identity
Human Mind Identity the Power and Intelligentsia itself
Creation: The Purpose of Progression enshrined in Book of Elements
Is to logically construct the five regular platonic solid state physics thus End of Syllogistic equals to (Progression) is to obtain Vision and the logical construction of these elements means the ability of human Intellectual system to construct fire, water, earth, air and light and what we are trying to tell and demonstrate and to prove is equal to a Massage;
Human Mind is the ultimate power the power that can create. And the context Message of this text book is the Theory of Creation

The world in which we live has been given an endless interpretation throughout the history of mankind but the most important obligation of science is not in definitions but in recognition and in changing of the world thus the core context of this Study is Book of Life that include Universe, History, Time, Man and the state of consciousness
The theory of creation and book of life Vision and astronomic
The Environment: Container
The Contained in the Environment Container Organism;
- Matter
- Plants
- Bio
- Animal Existing with no Faculty of Expression or Utterance
- Human In Possession of Faculty of Expression and Utterance
- Levity and gravity Ratio: Earth and Sky or High and Law Axioms

9. Regular Forms

Regular occurring instants, are detected by Human Mind and the Organ of human Mind
Recognises such occurrences as a Time (TYME)
Pentagon Regular Polyhedron or polygonal is Equal to the Form of Time Form of time the Container that contains Human Mind and the Faculty of Vision
Thus pentagon equals to a Universal Container that contains everything that Exist.
Pentagon is Equal to the Text or form and formula of the constructor or creator
The Ultimate Power
Pentagon is the form that contains the Elements
Tetrahedron, icosahedron, Cubic, Octahedron and Dodecahedron thus Pentagon is a form that every other true forms of Existence Emerges through
The process or progression is the Program of elements and the procession of Human Cognitive power from their Particularity to their Universalities or Generalisation

And that will be the method that we are going to present in this book then the readers will be provided with a tool to Understand Everything that Exist and then themselves Identify how far those topics are related to them.

10. Physical Interpretation of Below topics or revealing topic-jargons enveloped in mysticism of Speech.

Test these topics by your own five senses Topics:
1. Energy, General Relativity
2. Logo
3. Magnet & Electro Magnetic Field
4. Electricity
5. Radio Activity
6. Waves and particle
7. Sceptical
   a. The Duality Theorem:
   b. Relativity Theory:
   c. Wave or Particle Dilemma:
   d. Current Debate in the Domain of Science and Philosophy.

Theory of Light: Electricity and communication, observation and two dimensional (breathing) Systems

The Substance:
1. Substance Topic = 0|Zelot & Equal to **Letter** (no. 0 in the topics)
2. Logo Topic = 1|Text. Element. Letter Numeric (no.1 in the topics)
3. Numeric Identity = Char = Grammar of Equities & Law In Mathematics and equal to the Elmet's Topic of the Human **Consciousness** Placed on its **Sub- Consciousness** forming **quality magnitude : composition combined together in mutual relation Called circular Motion in an Equation and = Category = Faculty of Enumeration.**
Called Magnitudes and Common Notions, **in the book of Elements**

Magnitudes and Common Notions:

Statements:
1. In any Operation Elements Transform into Magnitudes and common Notions in Time

Element-Concept = Time in <u>Magnitudes</u> and their common <u>notion</u> Apprehension (Grasp)
And equal to <u>Strength</u> of <u>Gravitation</u> in the topics of Contemporary physics and called Right Angle in the Book of Elements.

2. Logos = text in contrast with Human Mind & Head and this Point is where The Text is at Right Angel <u>In</u> or <u>Of</u> the Text The Position Called Centre & Sphere, details of which Demonstrated in Book of Elements.

Please See Algorism & Schematic Logic, that follows;

11. Organun: the Instrument of Human Mind
Element of Human Consciousness
Prior to human consciousness first Exist
   a. Law and (System of Organic Structure) that Enumerate or Demonstrate thus to Enumerate is in Equal Ratio with to Demonstrate to Express to Say to Make an Assertion.
   b. **Categories Is** a Text from Organun that Enumerates Human Apprehension: Human Apprehension can be a Subject or Predicate for any Proposition. Five Axioms or Common Notions (Magnitudes and Common Notions) in a cycle of time the Pentagonal Regularity (Regular Form: Five Equal Edges and Sides and equal to Mass of the Cosmological Constant and are equals to the seats, Numerical Basis that Human Mind standing on our five senses are External objects **or constructed Means and they are in their Original Nature Mirror Organs, and are Constructed by the inner Feeling Axioms thus to Enumerate to Mirror Image this ratio Equal with Mirror and Light Interaction or Mutual Relation (thus common Notion are our five senses and that contains our feelings; Human Feeling.**

c. **Human Mind Authorises**, human cognitive power **Confirms and Not Deny** that those axiom and common notions to be imagined as pendulum or clocks or clock pendulum behaviour. **See Text Book** for these Operations and progression.
d. A, B a Pendulum Acts, as axiomatic Text that separates two points (magnet sides) thus equals to differentia that separates; In General: Day From Night Black From White. Since pendulum Motions Conducts both polls of a magnet or Human Hearing Medium in a space of a second.

12. Mathematic Logic

Human Power of Imagination
Arithmetic and Geometry: **Logical Means**
Music and Astronomic: **logical Mean**
Light: Eye and Sight
Topic Subject: Vision Electricity
CRT Cathode Rays and speed of light optics and Plain Geometrics
The method is to look at illustrations provided in the Text Book and imagine then to follow progression process and get guidance through persuasive rhetoric
Example Imagine: Arrows 12345 are Equal to Pendulum Magnitudes
1.2/2.3/3.4/5.5/5.1 are, Equal to pendulum touches (multiplications) in a Cycle of Time.
Each Pendulum Motion Acts as Seconds (1.2 Transforms to 2.3 transforms3.4 transforms 4.5 transforms 5.1 then transform back to 1.2 thus 1.2 completed its cycle (1.2: 23,34,45.51) Constructing a Pyramid compass the same for the other four five compasses
Construction of Time Compasses of time and time tables compasses: 1.2/2.3/3.4/4.5/5.1

Angle 1.2 Compass *Counts or|and Spins, Expresses, mirrors ...*   *2.3 3.4 4.5 5.1*
Angle 2.3 Compass *Counts or|and Spins, Expresses, mirrors...*   *3.4 4.5 5.1 1.2*
Angle 3.4 Compass *Counts or|and Spins, Expresses, mirrors ...*   *4.5 5.1 1.2 2.3*
Angle 4.5 Compass *Counts or|and Spins, Expresses, mirrors ...*   *5.1 1.2 2.3 3.4*
Angle 5.1 Compass *Counts or|and Spins, Expresses, mirrors ...*   *1.2 2.3 3.4 4.5*

Organun

Thus constructing pyramids: (P) first P1.2 then P2.3 then P3.4 then P4.5 then P5.1
12345 are equal to the Sequence of their Numeric occurrences, reoccurrences and their operations in time
Thus 1 transforms to 2 then 2 transforms to 3 then 3 transforms to 4 then 4 transforms
To 5 then the cycle repeats itself
The total amount of the operation in time = to demonstration times the organ Faculty of recognition in Human head which contains the faculty of confirmation and denial and it is always a rational referendum.
The subtotal of numeric substituted in the function of operation are equal to 5 numbers or angles time the five regular square bases are equal to icosahedron with 20 Angles or numbers and are equals to the container that contain the numeric or counting angles or numbers and the squares expressional means Mirrors expressing each particular angle and the physical reality of that is equal to the faculty of vision Contained in the Human Head Sight and Seen Sound and Hearing, Feelings Plus the Faculty of Expression:
Utterance, demonstration and observation in cycle of time and breathing system is in equal ratio with theatrics |sight to eye ears to sound times demonstration| the square of sight is also in equal ratio with the Market or the domain of political economy

1.2 Expresses (counts, Commensurate, Becomes Mirror for 2.3 3.4 4.5 5.1 and 2.3/3.4/4.5/5.1 Expresses (counts, Commensurate, Becomes Mirror for 1.2 in cycle of time
See the Text Book dictionary for the progression and the process of **Propagation of light in Empty Space[Vaco]** Contemplation through Illustration Elaborates Points as Text of Human Utterance as Human Five Senses (clocks as common Notions Sense or Mirror Imaging Process that takes place during the Multiplication Process (P1XP2) *see Illustrations*, Pendulum Senses
Thus everything that exist Equals to Pendulum or Clocks in time and the magnet entity is time in Clock and pendulum apprehension. And that is equal to Human Apprehension.

In time; we mean in eye of time, in the view or opinion, of time, relative to time and with reference to time.

Clocks Constructed, Pendulum Touches are Angles an angle is Clock of time or our Senses Constructed via Multiplication Process Carried out by Organun the Instrument of Time Is the Instrument of Human Logic and a container that contains Human Logic and Apprehension our senses are mirrors constructed to demonstrate our feeling and they are the Extremity and a boundary of our feel or sense of live Thus Prior analytics is the faculty of vision the power to see to hear and to touch to feel then to utter and express the power of imagination is the essence in other expression: <u>God The Absolute</u> can imagine itself. Thus the Question is Not in what we do not know the (?) Question is in Mistry (!) Mystery of life Contradiction and nun contradiction Factors, Who we are where we came from what is our true Identity (the Law of Identity) our existentialism is in Cycle so far no one know, where is the begin and end of it. and if there are any

Information regarding the creation and our End are merely Sceptics or Mystic.

Vision is our feelings that exist as a state or moment (In Us) tangible moments that we feel such as a passion, Pain Fire Cold Air Water and the Text Described in book of Element Equal to the forms of those inner feeling then our mind permit them to be accepted as true NOT **Deniable by Human Mind Recognition** and that is Equal to itself the text is the mirror via which the power of imagination See Feel and Hear Himself and construct propositions.

The Four Elements Fire, Water, Cube (Earth) Air are texts of our inner reality and feelings

To Burn to feel To take action to Operate and their texts are Fire = Tetra Hedron Water Icosahedron Earth Cubic Air Octahedron plus Dodecahedron the last one is the time that counts sixty seconds in a cycle of five seconds.

The progression of the text takes place in Empty space or Vacuum Tube this process is terminated by vocal expressions of propagation of light in empty space and construction of Vision to produce light, is in equal ratio with to reach enlightenment example to produce light is in equal ratio with to obtain enlightenment and that end cannot be achieved with out to have

Organun

an International Global Provision and Vison finally we reach a conclusion: modern times require, as a Historical Necessity, a Scientific Method and Universal law to emerge as, the historical progression of human history, which is right now At Globalisation

13. Text Book concerned with:

Optimal Solution of Inferential and Differential Equations and that entity is Organun the Title of our study, the Text Book Belo conscripts is to Identify the entity of Defention.

Manuscript.Context.Defention Equal to Equation:

Equal to Organun.Plus.EQUATION and|-

Zero.0.Defen.Equation.Rest.Plus.Compositin-Mind "Crucifixion" Divison.Subtruction. AND EQUAL to Organun.Plus and Medum-Human.Logic, Expressin.AT AND Equal To Vision.Plus.AT: Right Angle.Zero.Xs.Plus.At: IN.Circulation.Plus AND Equal to
INJN:- (**Of |OR**) the Circulation And Equal to Time. End
Proposition: Equal to A Pyramid Plus "The Right Angle Inscribed In A Circle" And Equal to
ENGIN: "In Jin. inCycled" "In a circuit" and Equal To Time, and "Equal to {The Conscript of EternityPlus.Clock} of time at Right Angles in Contrast with Zero.Xs and Equal to the Angle of Tita and Equal;- (A 3 Dimensional Object IN our Reality)|- Greek Name; Tri Via Incubated In
A Solid State Physics|- Called Cubic Plus, Right.Angle.at (of or In the (Cubic Container) in Time.
Thus:
The Time Sensor Equal *to* a Proposition that states:- Right Now: Time and WHAT is a Time (?) the Quest- tion and Equal to Mass Times Speed of Light. and Equal to Energy Conserved Clock
The Sceptical Equal to Tri Via
State Of Uncertainty and Equal To Electric Charge That Demonstrates Right Now {In CRT and Cathode Rays Operating system language

Function} <u>That Called Computer-Cursor, Controlled via Keyboard or Mouth</u>

<u>Proposition;</u> In the Subject and Of the subject
Anything that Exist Equal "Magnitudes or Common Notion" In Time Sense of Time = Time in Magnitudes Plus Common Notion Apprehension. End
And that equals to Faculty of Vision
Human Logic of Utterance "Elim Al ManTique"

Islamic translation of Organun, book of elements and transition of the Text:
Organun Translated as (Container that contains faculty of Vision, the science of Expression, Speech and utterance) Kitab Al Mantiqe, EssaGogy, for Exegeses QataGorias for categories Falsafa Al Ula and Al Falsafa Althanya, first and second philosophy, for prior and post analytics Topos al-Taaliqat, *wrong qiass or Scale* for sophistic refutations IBNSINA, History of Ibn-Khaldun, Farabi, Omer Khayam Ruaiat, for Quadrature and Newroz Nama the new day massage. Messages of Safa Brothers, Ibn Rushed, Islamic Sufism. Assassins and Ismailia movements
None of Muslim Scientist or diaspora involved in translation and transmission of the text was Arabs: were Iranians; people who are known by Kurds are original spices of today's Persian Iran and they were Sasanians the Empire ruined in the emerge of Islam.

   14. Further Illustrations:
0. Context of the Book
1. Book Title: Organun
a. Brief History of the Title Organun
b. Organun the instrument of time.
c. Organun The Instrument of Human Mind
d. Organun the Logic of Human Mind
e. Human Brain: Mind the ultimate Power
f. Power of Imagination

g. Human being The Entity that is in the possession of (ultimate power)

15. Context Contains:

> An Invention Tilted <u>under</u> one. Single Topic = Organun That MEANs an
> Instrument = TIME Which OPERATORS via An ENGINE (In-Jin) Called Radio
> Activity. RADIO-ACTIVITY contains PLAIN Circle OR Circles in mutual relation all are Operating In a single organic structure in Contrast with Human Logic (with reference to) - Hunan Mind, Hereafter: Faculty of Vision **In Context of time. (Universal TYME) for:** ORGANUN Method **of Construction Using: Boolean algebra, the Element, Operation of the elements in a permanent cycle of in time: Illustration on the essence of Binary-system Method of Contrast:**

Of design Chemistry Electricity, Current ac dc Current: the Organun contribution is in Propagation of Light in Vacuumed Container, CRT, and Computer, TV, Screen and Radio waves, Or in Empty Space. Cathode Rays, X-rays Laser Beam Spectrum these Products are all from the current which is produced Using Magnet and Induction: electricity production Via Induction. Thus the study is concerned with: Topics:

Theory of Light in contrast with, Parallel with Theory of Time (Universal/ Time) And Organun: **Universal System of Human Intelligence** known by (the Essence, Origin Human thought Episteme, Epistemology, Faculties: The Abilities....) The Essene (0) Zero and its Categories. (0,1,2,3,4,5,6,7,8,9) Ten Numbers

15. Brief History of Book of Elements: Pythagorean propositions
- Politics
- Golden tune
- Music
- Theory of numbers
- Metaphysics translation and interpretation means
- The Angle inscribed in a semicircle is a right Angle
- In a right angle triangle
- The circle and semicircle
- The diameter of thee circle
- An arch, arrow or arrows of time.
- Book of Elements in <u>contrast with</u> Pythagorean Universal Theory or Tallies, Tri via and Quadrature. Squares via method and provision used in antiquity, in teaching and application of the holy bible, also a method in the book of elements tri via (Pythagorean/ Tallies: AL TALUTH Hebrew Arabic.) 3 Dimensional objective Reality
- Time Place Intersect Tri Angles and time Motion| **Spaces Intersection** squares ....

Method of Design a DICTIONAY / Golden Tune: tune to radioactive median via an optical means Crystal Logic and alloyed material that can convert cosmic rays and radioactive, radio, Wave Particle Messages as a source of energy, feeling inspiration and inductive methods.

The Organun Text Book is a Product of a Self-Research
a. On an International Scale (self-experience as International Director on Humanitarian Issues)
   Covering Domains that Contain immortal topics in the history of human being recorded in time
b. Theology
c. Anthropology
d. Political Economy Classical Economics Smith & Ricardo, Austrian School of Economics and Keynesian theory of Employment Money & Circulation
e. London School of Economics and oriental Studies

f. Metaphysics, Physics, Biochemistry (Science of Magic) and Oriental Philosophy
Ancient Testimony, Holy Bible
El-QURAN and Exegeses or Science of Music, Classical physics Galileo classical astronomical System and the law of Inertia

g. German Philosophy: German Astronomic Mystery of Cosmography Hegel. Kant Marx Theory of Value and Dialectics
Quantum Mechanics
Duality Theory Relativity| Einstein's earlier papers Light & Wave Particles Duality: Einstein's Universal of Equation Mass=Energy X Square of Light, The Sceptical [Current debate in (domains of Science & Philosophy]

h. Brief **History of Time**
The Question or the of Proposition [*Determination of Cosmological Constant and Strength of Gravity*] is equal to the Conquest and victory of Scientific Knowledge, and that is in itself equal to Salvation, Liberation of Humanity and the fulfilment of the ultimate Human Ambition, Doom of Fetishism, alienation and Capitalism thus emerging the sun of Globalism, Internationalism and the pending State of Perpetual Peace (Al Islam/ Quranic Postulation, Islamic State/ Arabic Translation) and withering away the State of Roman Values Build on infidelity and Slavery Currently Represented by Western Imperialist Ideologies.

Steven Hawking: Department of theoretical sciences and applied mathematics Cambridge University.

i. Principia Theory of Probability Newton/ Galileo classical vision.

Guide to the readers of the text book

Please find the Text, Syllogistic and Scanned Illustrations, Elaboration and further simplified details that assist the reader; personally imagine how Organun Proposal Works and how individuals and Industry can produce Such an Instrument to obtain Vision and Instantiations, simultaneous transmission, Translation & Interpretation.

And how this Method involved and applied in the process of Meditation and Self skill Extraction Obtaining Qualifications via tunings to the text and following the progression **in contrast to their individual minds Via the Operation of the Axioms and Common Notions; their own apprehensions** and their own culture backgrounds which is the foundation of their own philosophy; their own arithmetic logical

The way is naturally their minds tuned to their own Stars

Plus how individuals and groups obtain capability to express their own vision and how they are becoming in the possession of their own methods of contemplation in to inferential and deferential realities thus capability to produce out of their own arts naturally planted in them out of the abundant in nature. Man and the elements are contained in an Environment, earth, air, water, fire, light Plus the Faculty of Prediction which is planted in Human Heart as a desire and ambition which is is hold in the future.
Every Individual being contains a value and the Algebra; The Obligation of Life; the Breathing system in which the organic being is extant the faculty of Human mind pressurises human Physiological system to extract out of his own value because it is the only way that can fulfils Individualistic Ambitions and that is in itself the Cause of human probe and human suffering.

You are Ether a Slave Labourer or an Affluent and Liberated
Having not your own vision
You have the first option
Obtaining Vision
Is getting
The Second option
The
Organun Text Book
Is equal to
Your guide

ON THE ORGANUN DICTIONARY

a. Vocal-Sight, Seen & Hearing are in Equal Ratio with to read from a Text and that is in Equal Ratio with to; Eye and Visibility [visible or Invisible Vague or Clear the form is rational not rational nice not nice] are confirmed or denied by the faculty of vision thus Eye possesses a faculty that can judge for instance knows that is a square and that is not a triangle thus distinguishes; Separates a from b; black from white. And that is called Contrast faculty the ability to Recognise and that in itself equals to the Faculty of Recognition

    a. The Conclusion is that; Vision is equal to Eye and light is its text
Thus the Element of Human Brain, Senses and Psyches are Equal to Physical Reality of Light
    b. Dodecahedron is equal to (Text Form, Formula) of Light and
It is equal to the instrument of time
The logic of human mind
Human apprehension
And
Human biometric Counts 60 per revolution or a cycle

Syllogistic
The progression of the text

Syllogistic is equal to [Progression of Human Cognitive Power from Particular to Universal] and that is in itself is in equal ratio with the History of creation
The Sun Rise and the reason for its progression is to find a Destination; a Resolution

The sun rise to sun set is in equal ratio with the emerge of light and its progression to its last destination the Human Faculty of Vision The Eye thus Eye is Equal to Darkness a black hole, and the boundary of light, a state called Night in which light get concealed

In Optical forms
See Dilemma, Quanta theories on Black holes, speed of light and its other fictions

The
Quanta mechanical propositions; are
No more than infidel fictions or weird poetics

The nature of human faculty of vision

The human faculty of vision is equal to light and the Element of light is a resolution performed by human senses

The nature of our senses

Is Equal to Mirrors and are in mirror imaging operation and procession and are equal to the boundaries of the cycle; the {Time the permanent universal cycle of existence}
Identification and recognition of the nature of our senses is equal to the definition of the cycle and that equals to Quadrature of the circle The Spanish Miracle
QUADRATURA Del CIRCILO and equals to the Element of the {definition axioms and common notion} See book of Elements

The non-recognised, not known Fact in the Book
of Elements and Aristotelian logical works
Thus equals to the limitation of
Euclidian Geometrics
Aristotelian logic
Greek Philosophy
And the Western Thought
Its

Institution
Sceptical Science
Weird Art and Philosophy

These Principles; {definition axioms and common notion} are indicated by modern commentators and they are in themselves equal to the expression of the Prior Analytics

Thus the posturer analytics is the faculty of mathematics, pure mathematics contained in the human mind and its probe is to reach its final end the fulfilment of human ambition the self-recognition and self determination

The self-recognition

In itself is the conquest, obtaining vision and the day of Resurrection.

Grey Level Marks; Black and White Intermediator

Grey level points or spots are organ senses with the faculty of Recognition; the sense of the deferential that separate black from and that is the analyser Magnitude which is carrying out axiomatic functions and is equal to the Senses of black and white

The Physical Reality
Of the cycle
The Time
The Cycle = Mass and Volume of Human Consciousness
Cycle contains:
Time = Text of Human Apprehension and equals to
a star called Pentagon = to the text of time which is a
composition pendulum or clocks with reference to
The human mind and
Clocks and spins with reference to human mind and its boundary
And

Equals to
The Referendum of human mind and

The Instrument of Human Consciousness; the Intelligence the IN-Gin and that is called philosophy

Mass No. of the cycle = five axioms or common notions = 5
See the text book
Five Numbers and Five Touches
The touch positions are situated in between of the numbers
They are equal to time
1x2 the 2x3 then 3x4 then 4x5 then 5x1
Then repeating itself
1, 2, 3, 4, 5 are = to Pendulum Magnitudes
And the composite numbers 12 23 34 45 51 are = to touches
The
Multiplication process takes place in the cycle of time
And
Registers
Five regular occurring instances which are recording a
Stare shape called
Pentagon
And that is equal to the logical instrument of our senses
That resolute the light and Equals to Light
Se the Progression of the text

{Regular occurring instants, are detected by Human Mind and the Organ of human Mind

Recognises such occurrences as a Time (TYME)

Pentagon Regular Polyhedron or polygonal is Equal to the Form of Time Form of time the Container that contains Human Mind and the Faculty of Vision

Thus pentagon equals to a Universal Container that contains everything that Exist.

Pentagon is Equal to the Text form and formula of the constructor or creator

The Ultimate Power

Pentagon is the form that contains the Elements

Tetrahedron, icosahedron, Cubic, Octahedron and Dodecahedron thus Pentagon is a form that every other true forms of Existence Emerges through}

And are called tangents the tangible moments in time and are equal to the units of our Senses and the physical reality of them is equal to <u>piece of magnet that contains a magmatic Field that the human head is positioned in side</u>

Radio
And the Radio Activity
Radiation and Receiving Radiation is equal to the function of Organun; the Organic being; the breath taker and the Matter Function is echoing
Thus Human Mind is acting like a receiver
That receives Waves and Also Send Waves
Time is in equal ratio with the Human Heart and Heart Beats
Thus Human Apprehension the Numeric bases Science is equal to Heart beat numbers
In the cycle of time which is a limit that equals to one minute

The axioms or common notions are the Element
Entity of Human Brain; the human
Faculty of imagination constructs forms for the (axioms or common notions) to transform them to be visualised by the human faculty of vision and those Symbols are equal to the text of light
$5*5=25 -5$ (substituted in operation) and that constructs a Pentagon
Pentagon = Text (inscribed in the cycle) thus
mass of the cycle = mass and volume
5 = Mass no. and 20 equals to Vol. No.
The Vol. Element = Water
Icosahedron = Text Form Formula of the water element and
Pentagon = Fire Element is a physical reality
of the thinking element and
Human intellect and that is the one in the possession of the faculty of demonstration and Expression thus the process of creation equals to
Self-Manifestation

Farek J Sadon

> The Instrument is the mirror and the self
> Watches and express itself via that Magical instrument

Human Feelings Natural Feelings Such as Passion and Pain and are equal to numeric square roots inscribed in forms called **quadrature, quadric is a quarter of a square** and equals to time space intersects, are called **clocks and spins** thus quadrature equals to a Pyramid and Pyramid is equal to **(structure or composition in time) (Pyramid apprehension equals to Time in Clocks and spin** *Grip, Understanding or apprehension*) Structures are [Weights- Grammar entities] defined as Tension, momentum, gravitation… and they are equal to indexes of Human Feelings. A **Pyramid** equals to Optimal Equation and equals to the Cubic Equilibrium. Squares are Prototype Forms of Human Feeling and are equal to four Numbers [4X4 = 16 – 4 = 12/4 = to a polyhedron (geometric forms) with four Equal faces, each of the faces is equal to the expression of the other 3 Numbers by one of the Numbers. the four numbers substituted in Arithmetic-Geometric Operations construct a compass in the form of **tetra hedron Water Volume = 20 Tetrahedron, HEDRA constructs 20 tetra Hedra compasses**
Water Cycle contains 5 regular squares constructed by the pentagon mass numbers
Each square Equal to Clocks or Spins in time
Time always equal to a Pyramid
Thus the cosmological constant = Pyramid and that is equal to ¼ Ratio of the total Vol. of human apprehension
20 tetrahedron divide by 4 = 4 * 5 = pentagon with 5 square each square contains 4 tetrahedron = Text of human eyes and hearing media and equals to faculty of Vision

Organun-Text. Programing Language

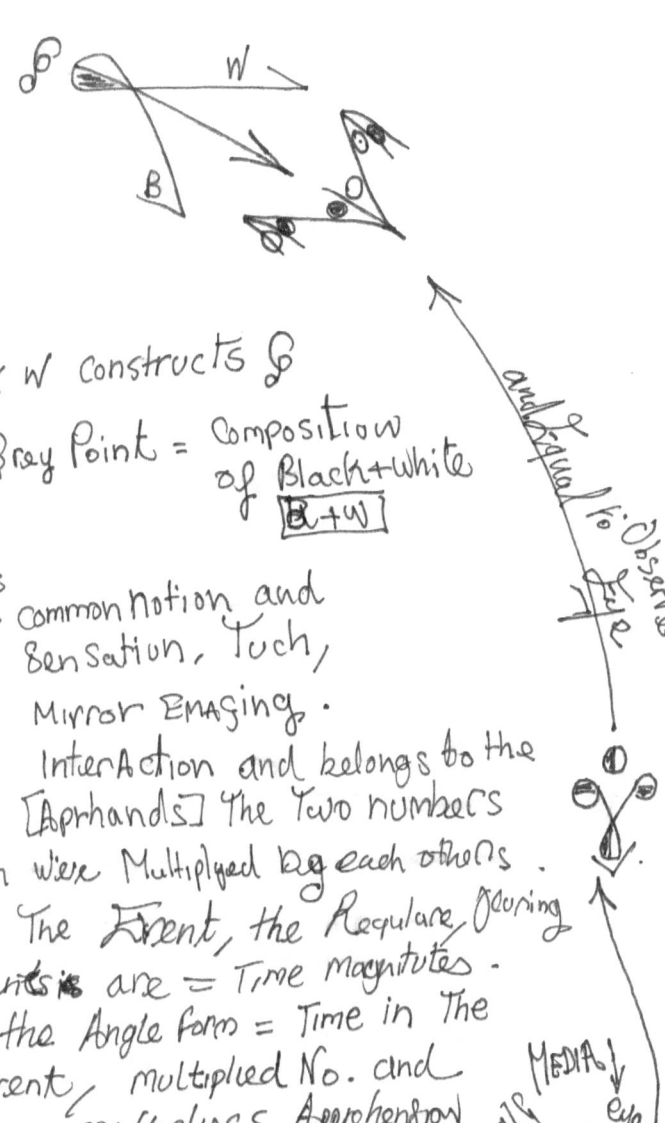

B × W constructs ɢ

ɢ = Grey Point = Composition of Black+White [B+W]

Thus
ɢ = common notion and Sensation, Touch, Mirror Emaging.
Interaction and belongs to the [Aprhands] The two numbers which were multiplied by each other.
Thus The Event, the Regular Ocuring Instants is are = Time magnitutes.
and the Angle form = Time in the Event, multiplied No. and multipliers Apprehension on that Rears to HUMAN FACE: HEARING MEDIA eye EARS NOSE

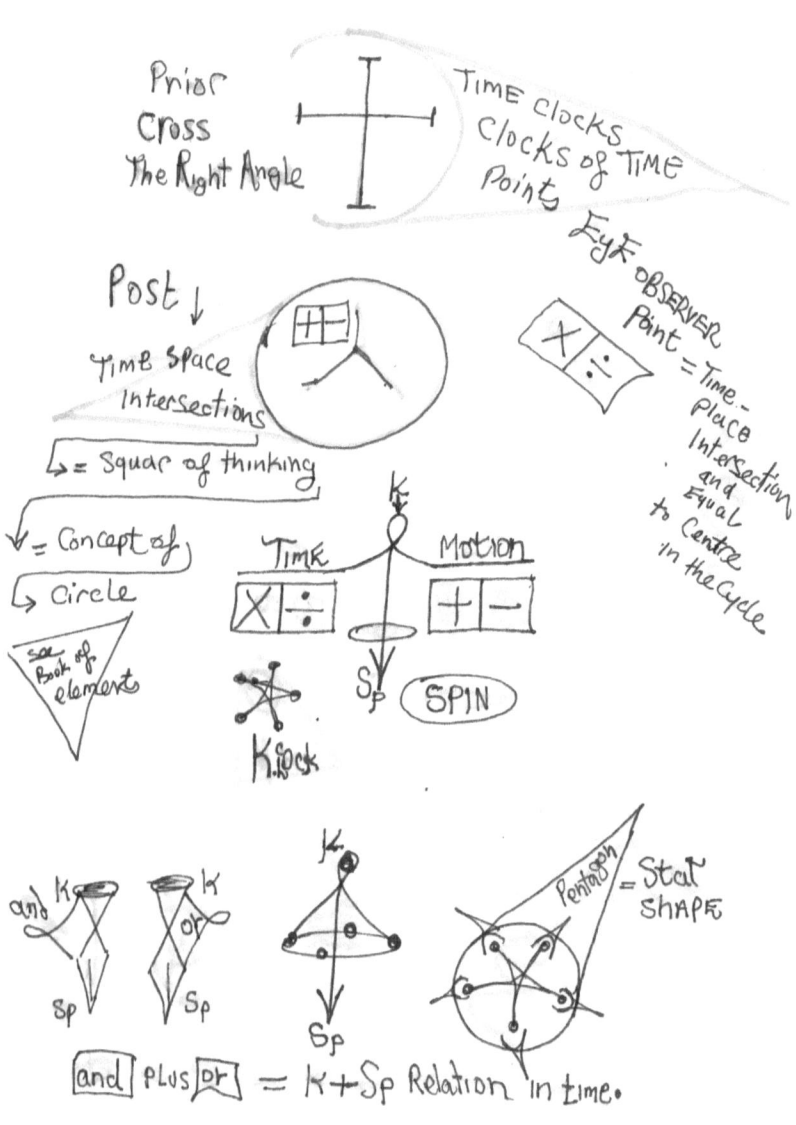

|1|2|3|4|5| are Equal to Pendulams

|12|23|34|45|51| = K: clocks

= Time in Structure
 and compostion
 Apprehension
= HUMAN Face.

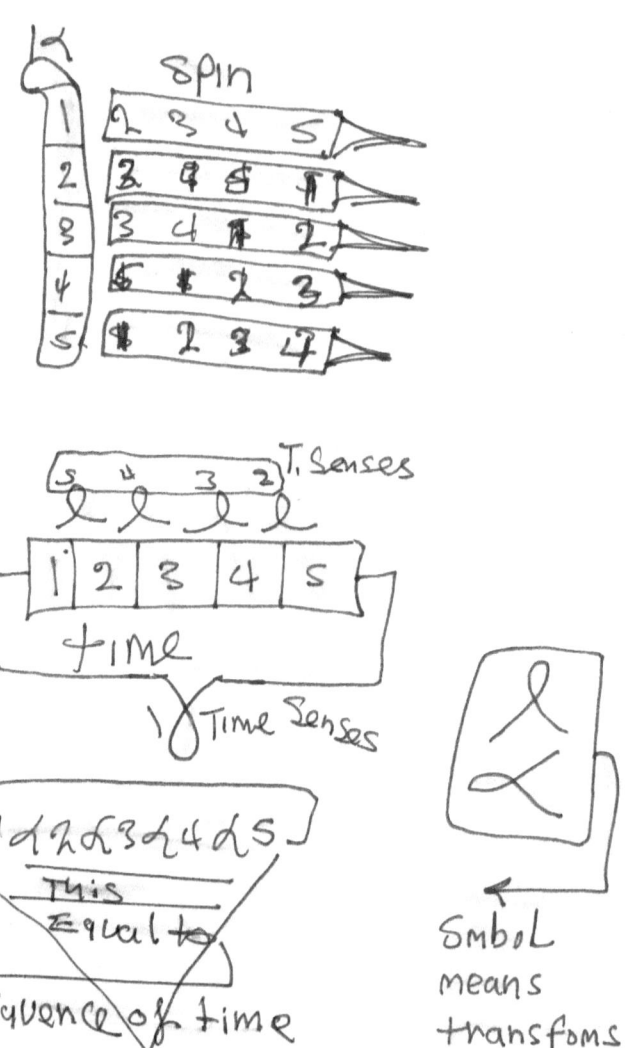

Smbol means transfoms

5 Numbers and their Transformation in time Thus Time = Transition Period. and EQUALS To Homan Application

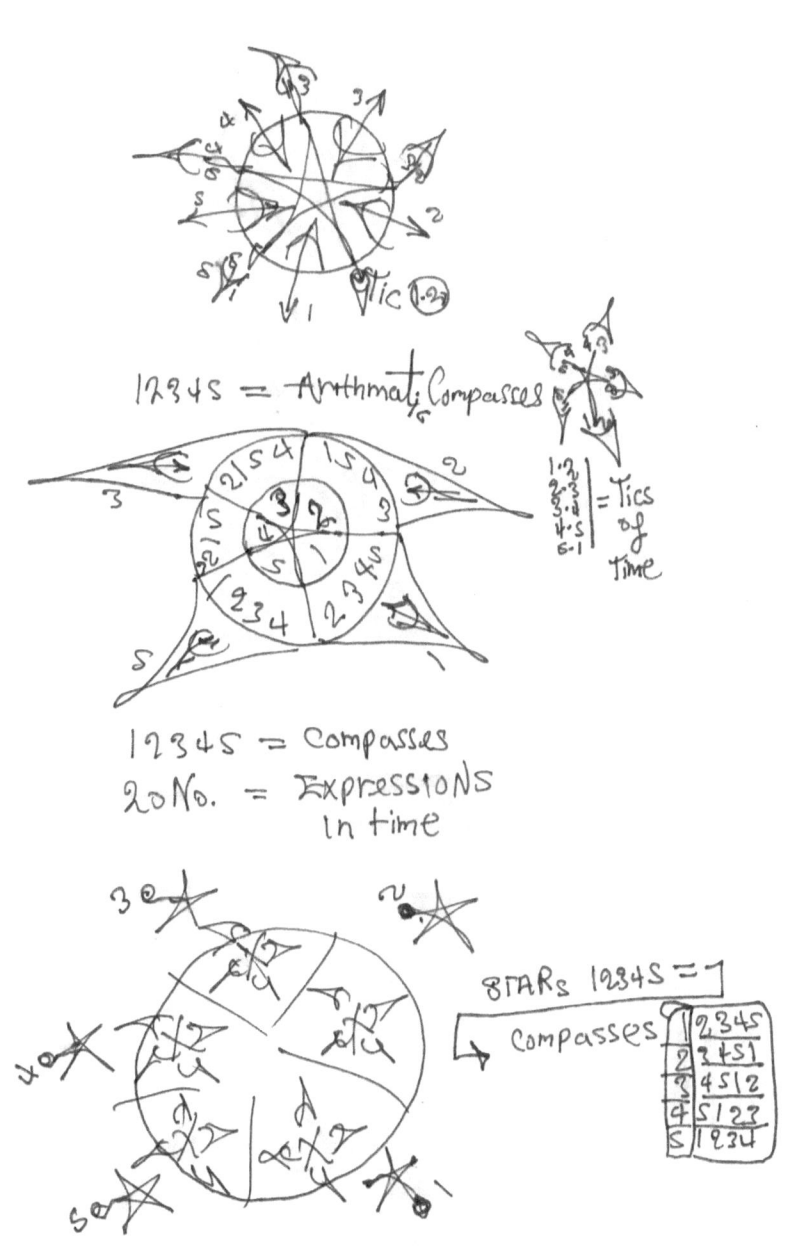

Thus in the Cycle
5 E^p's 20 No.
AND
5 are
Expressed in
four ☐ squares

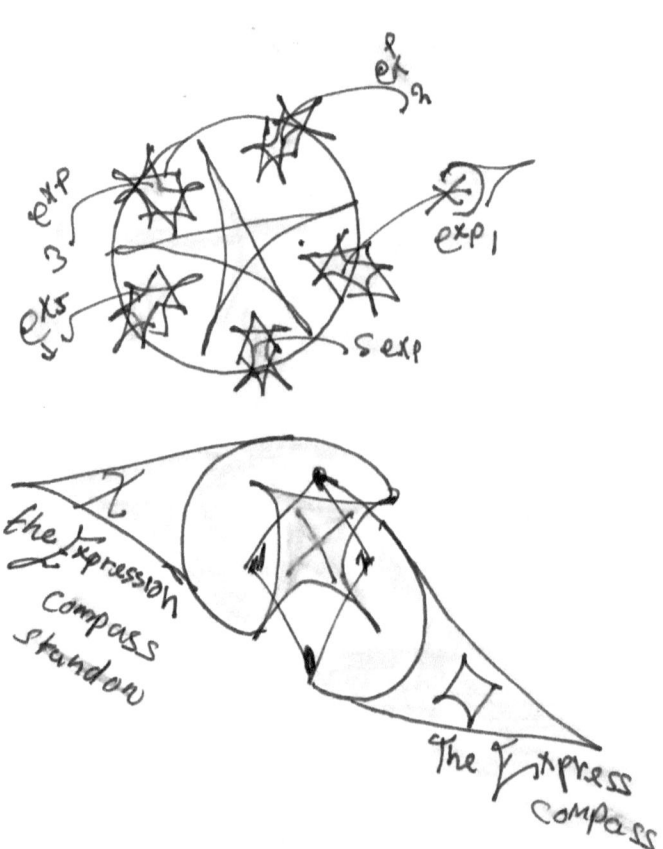

the Expression
compass
shandow

the Express
compass

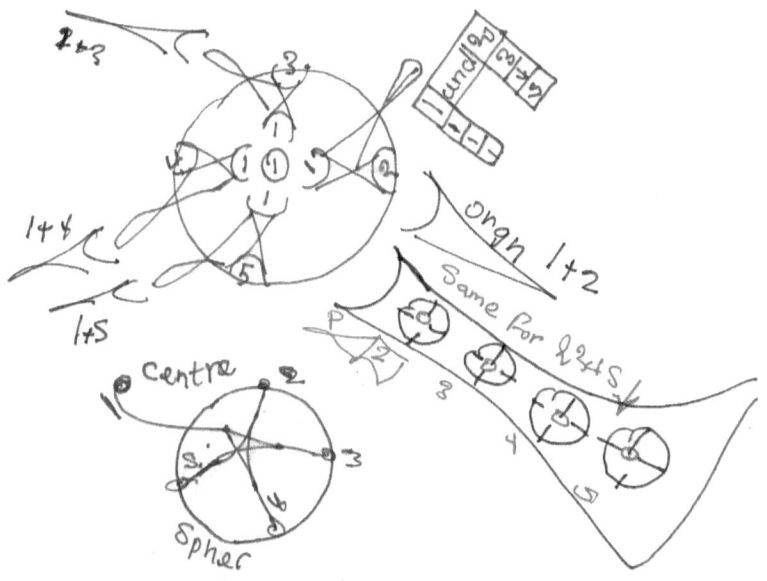

Apprehension Equal to
Center in K_1
Sphere in K_s 2345

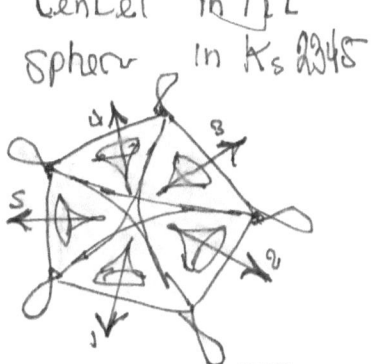

12345 are Equal to ✗ compasses Exp'ring 20 No.s in one minute and constructs 5 X_s which are equal to Expressions [e^{xp}] of the K_s

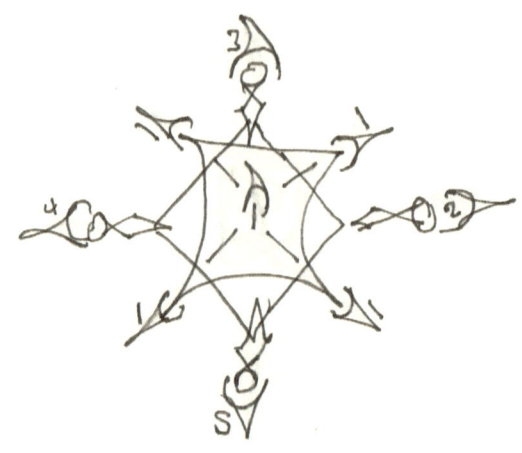

T = Spin at 2345
$T = K$ at 1
Pyramid Form = shape of $\boxed{K_1}$

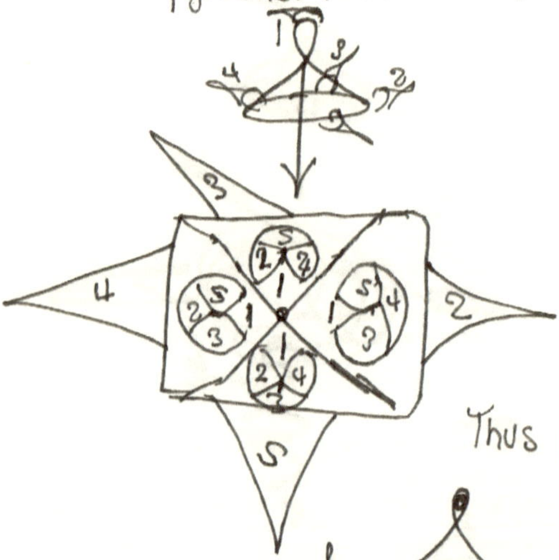

↓

Expression of K_1 in
THE Mirror of
↳ 2·3·4·5 K_S

Thus $\boxed{1|2|3|4|5}$ = Apprehension of The Pyramid

K $Sp[2\ 3\ 4\ 5]$

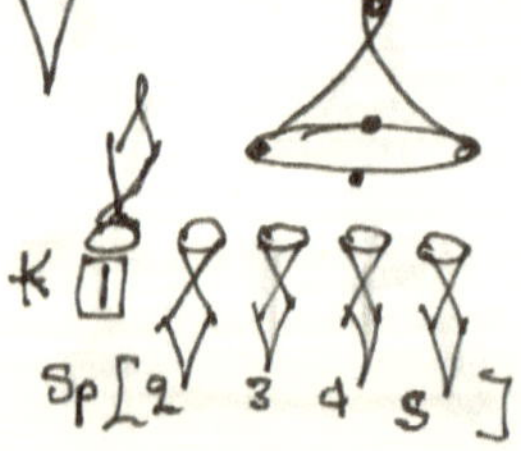

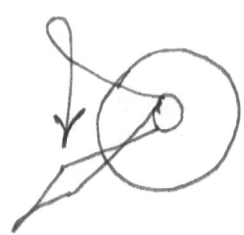

1. Equals to $\boxed{K}$ or $\boxed{Sp}$ in time
2. Equals to $\boxed{T}$ in $\boxed{or}$ plus $\boxed{and}$
   Apprehention or grup.

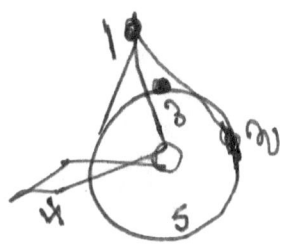

time = K in 4
     = Spin in 2&4&5

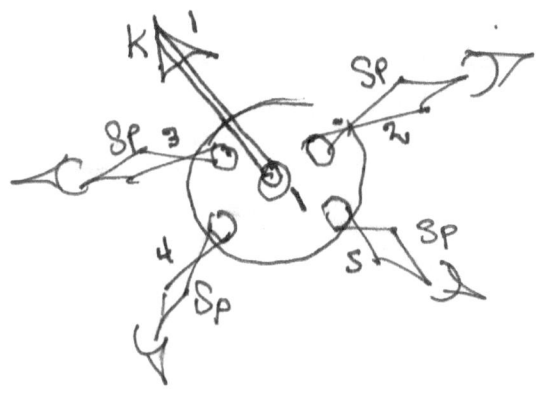

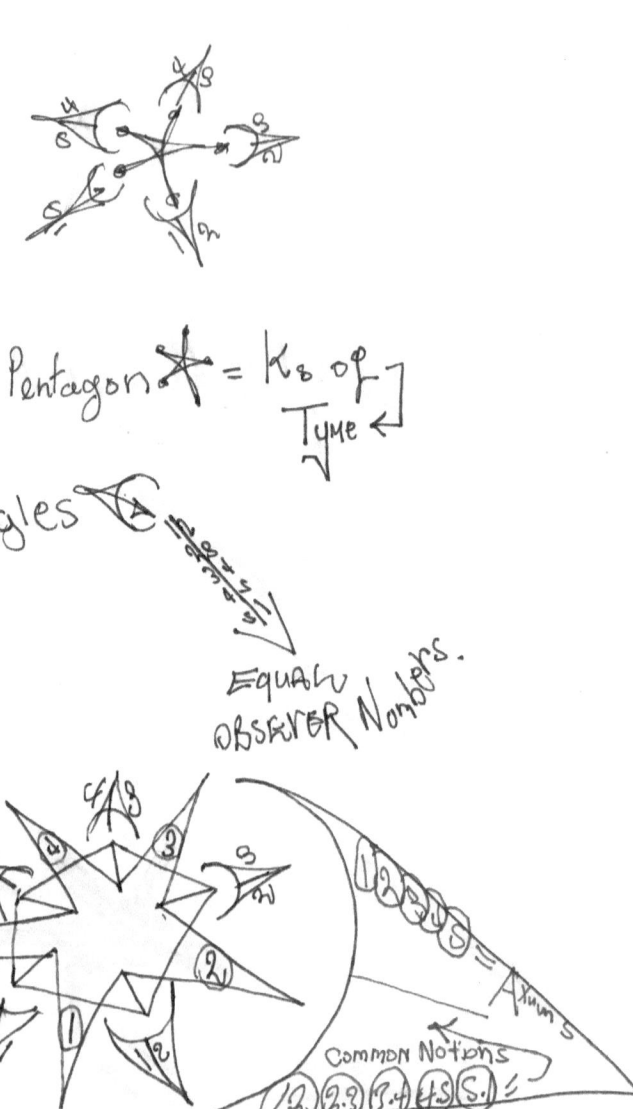

The Pentagon ✶ = K₈ of Tyme

Angles → Equal Observer Numbers.

AXUMS = HEARING + SEEN MEDOM
         EARS + EYES
Yrs of Time = MAGNITUDES of Sound
    Vocals Detected by Mind

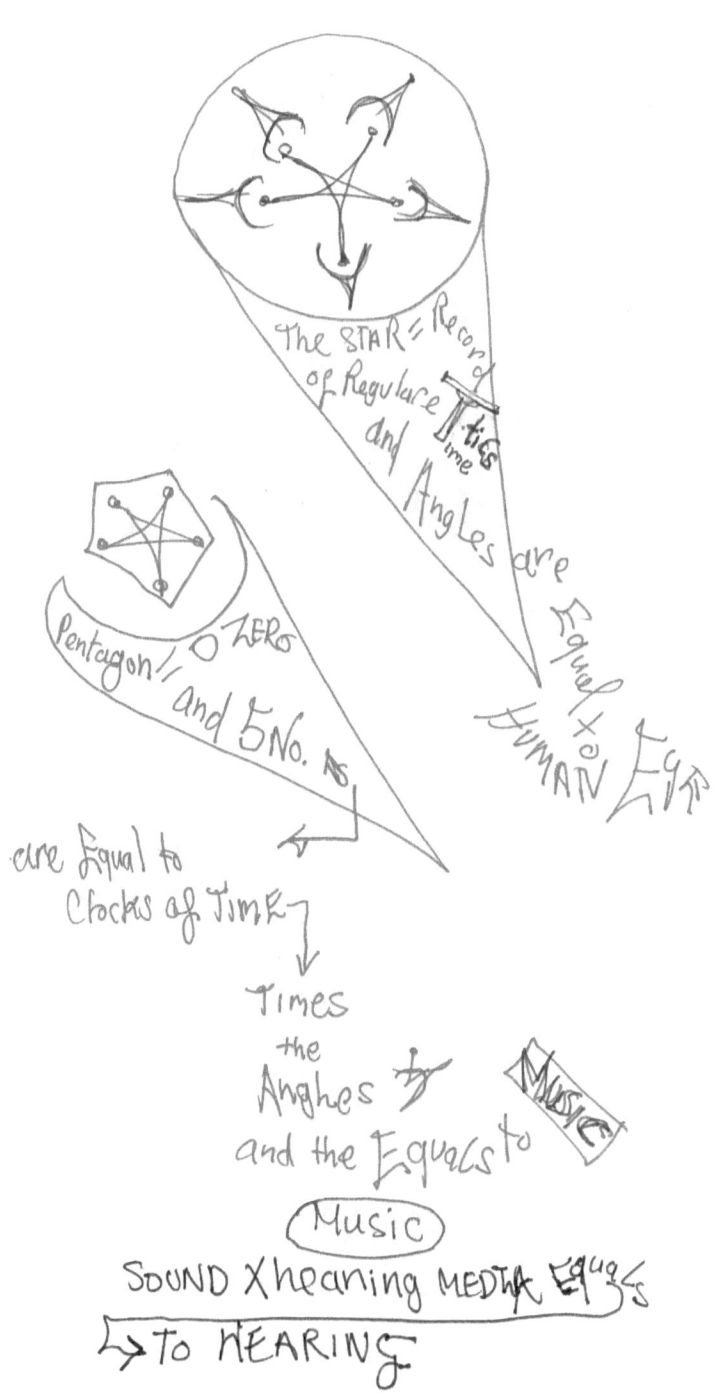

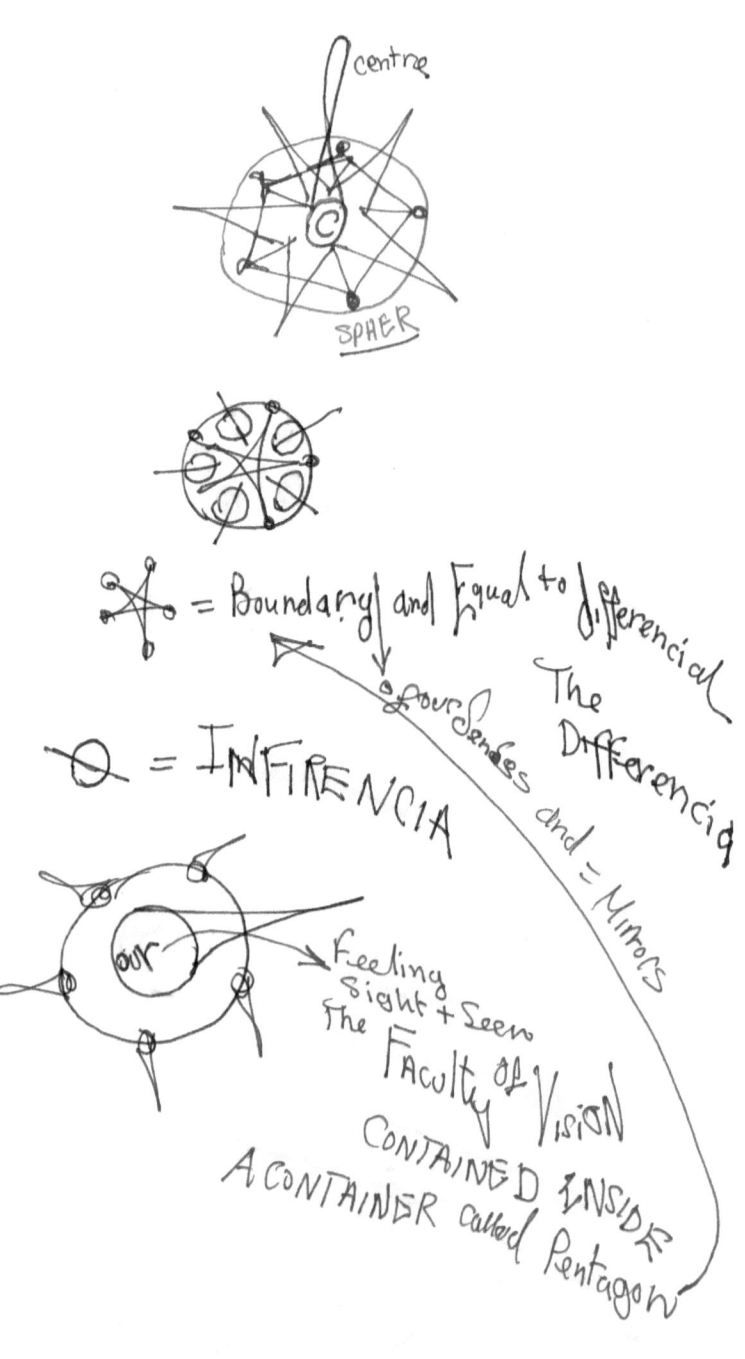

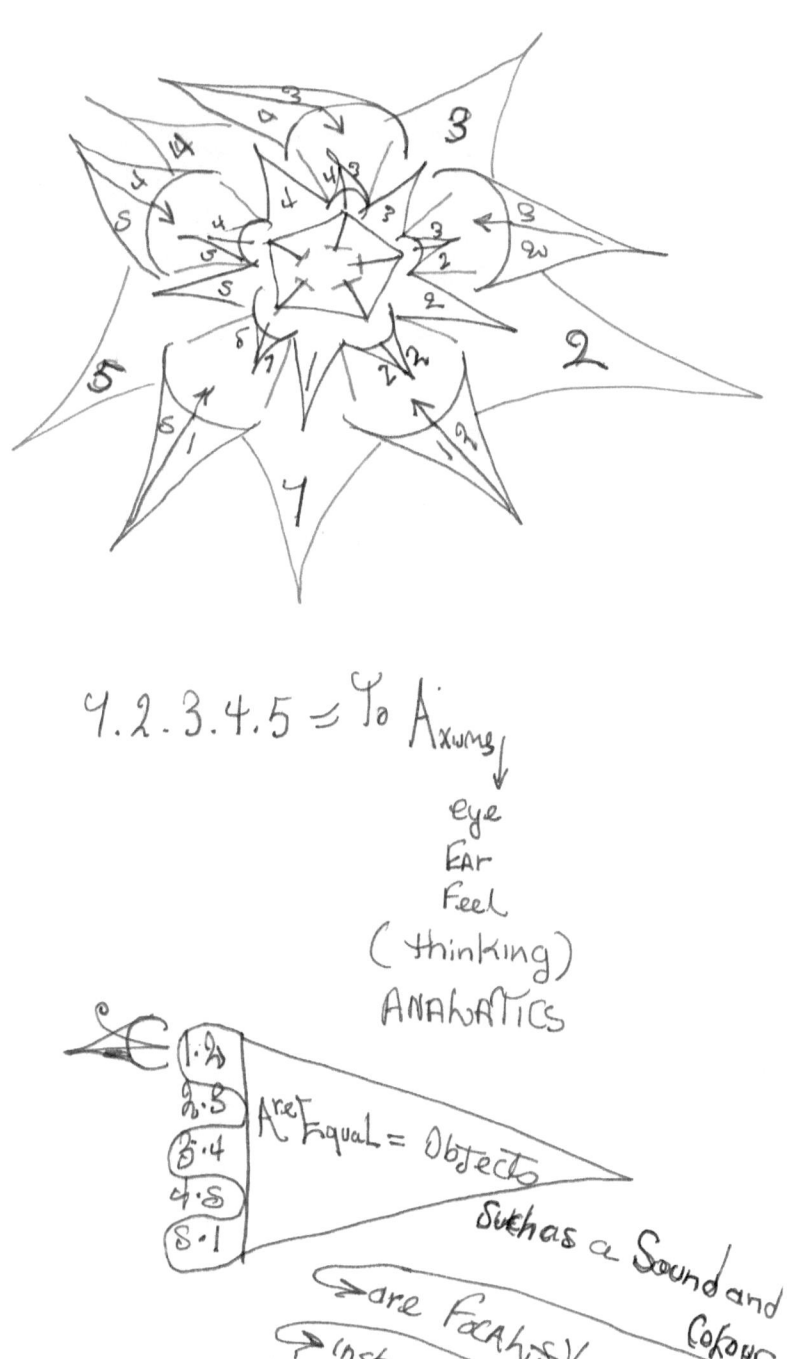

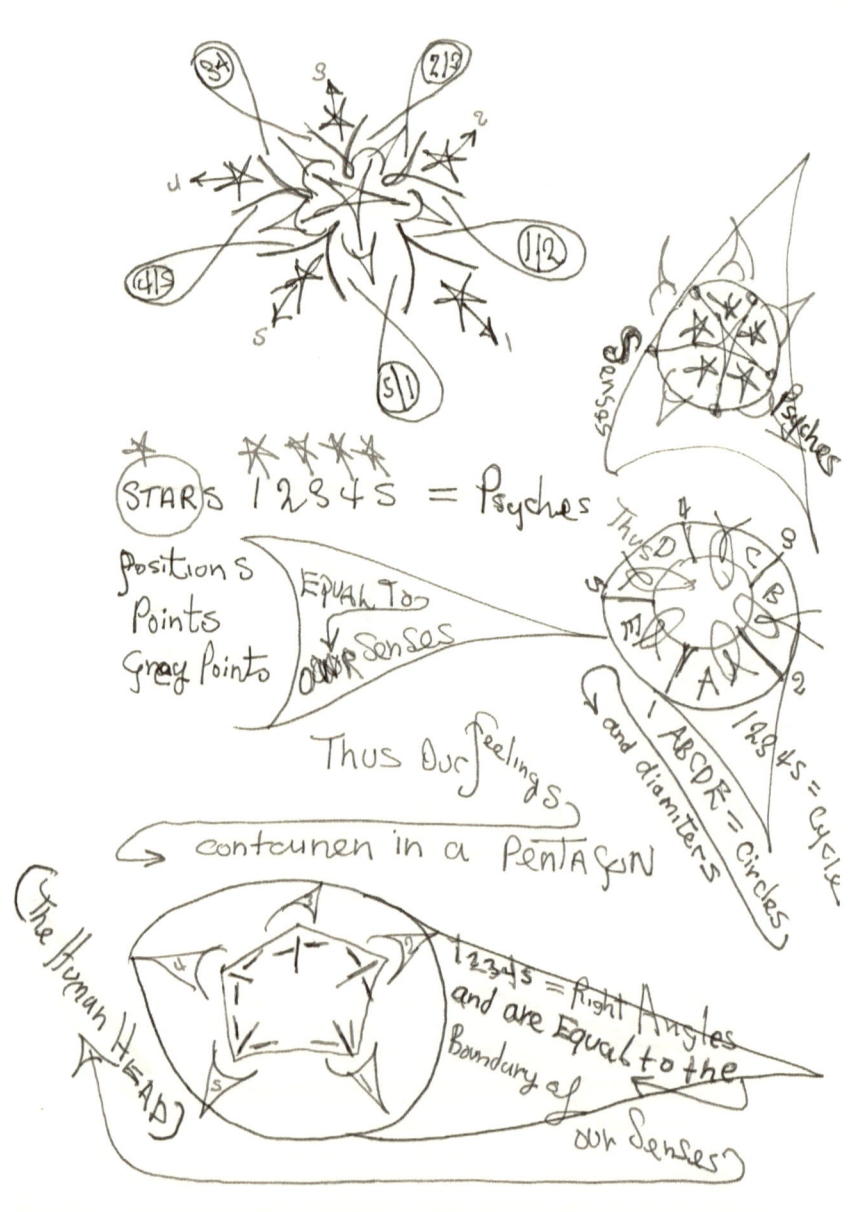

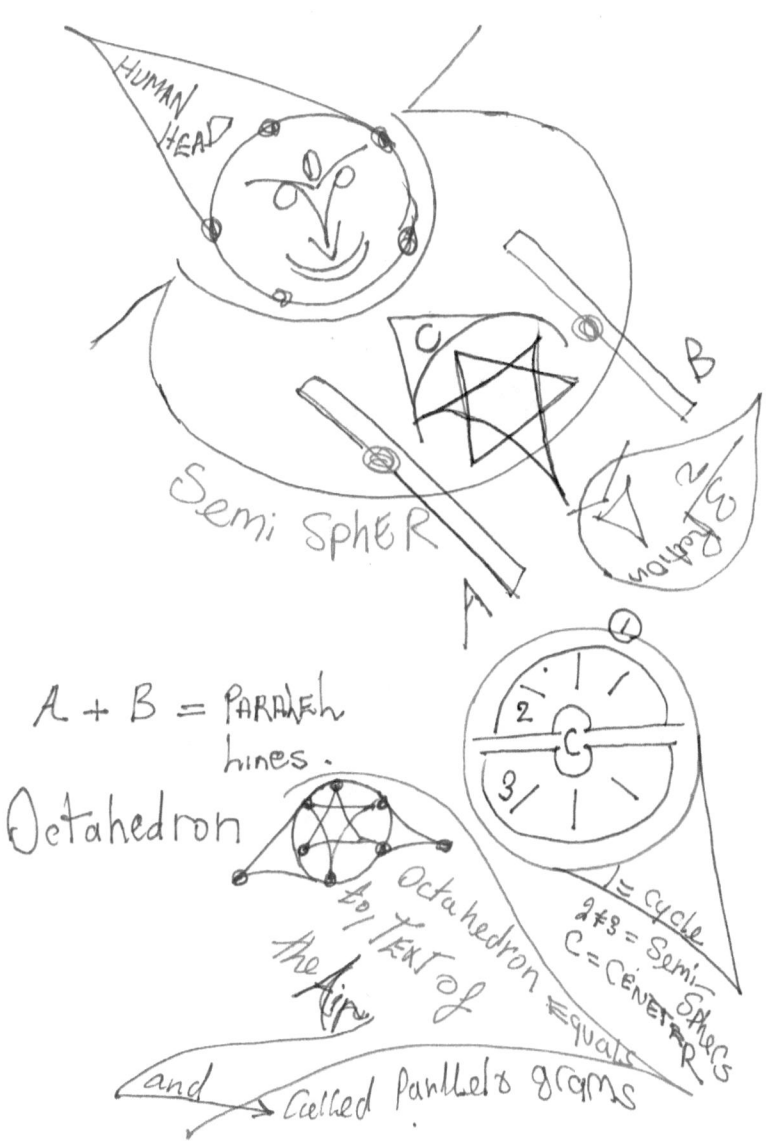

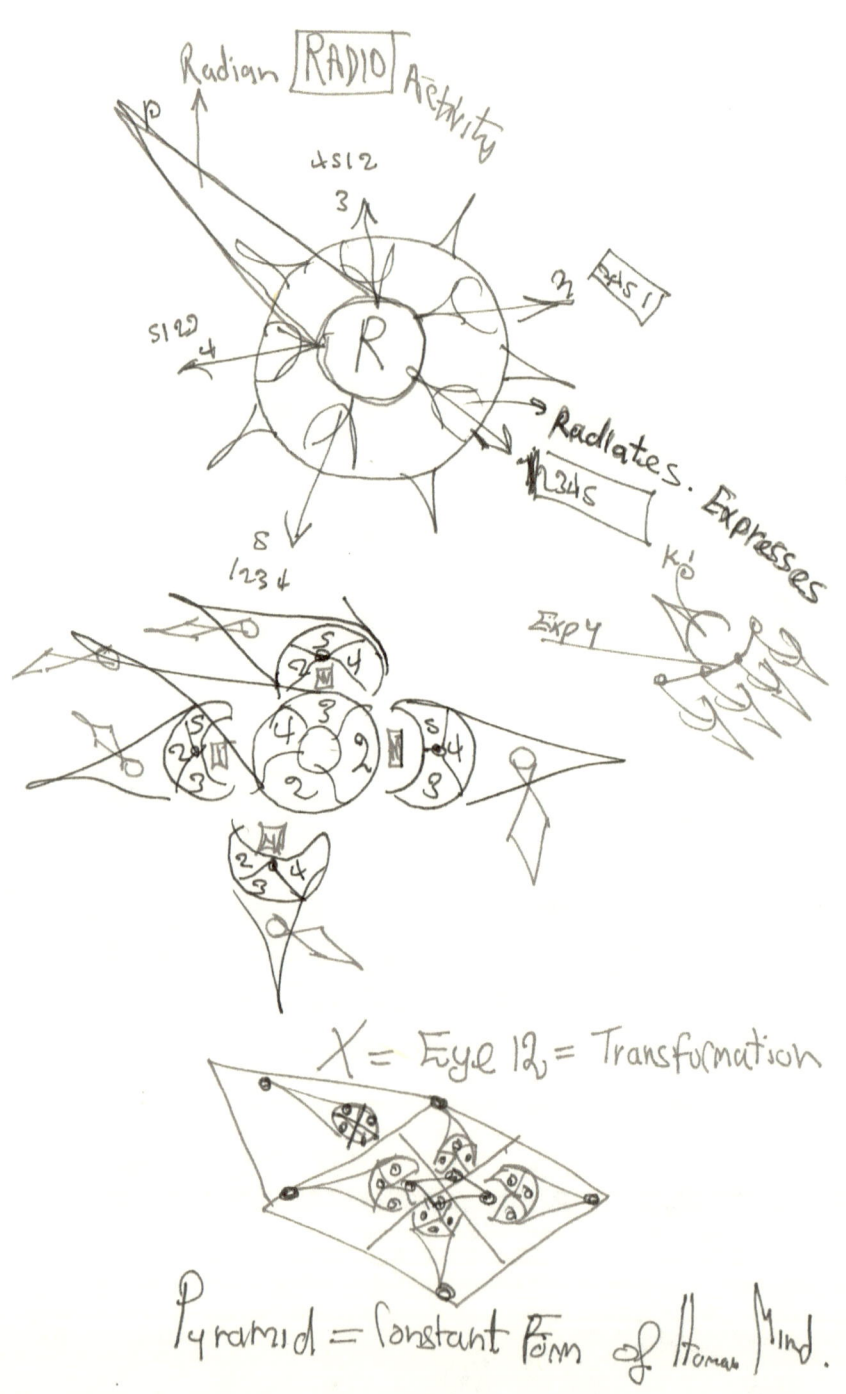

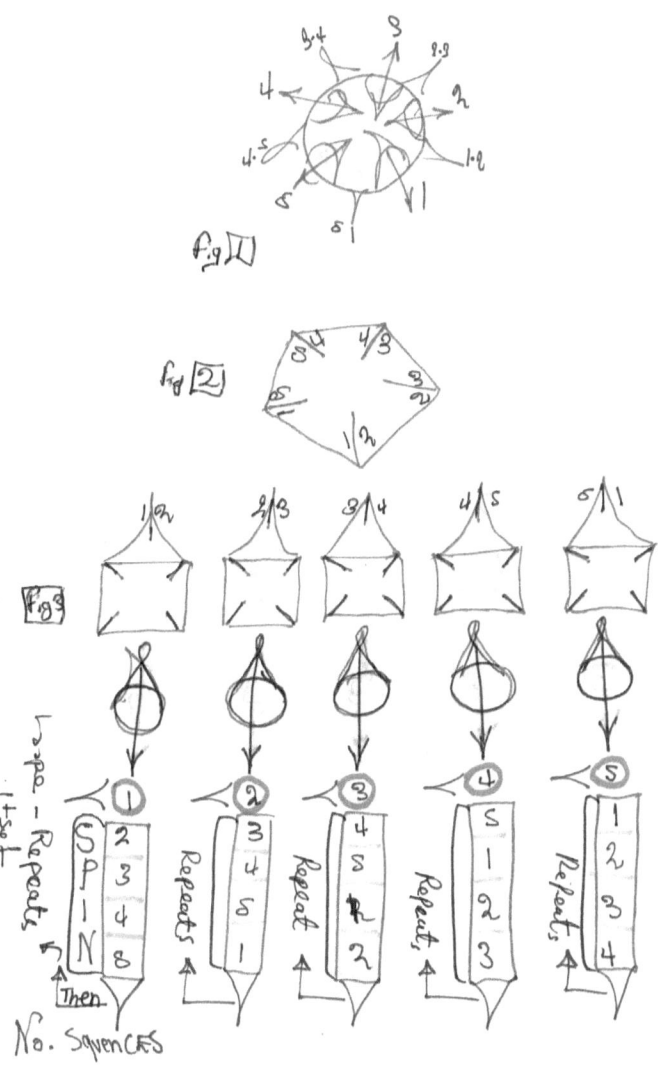

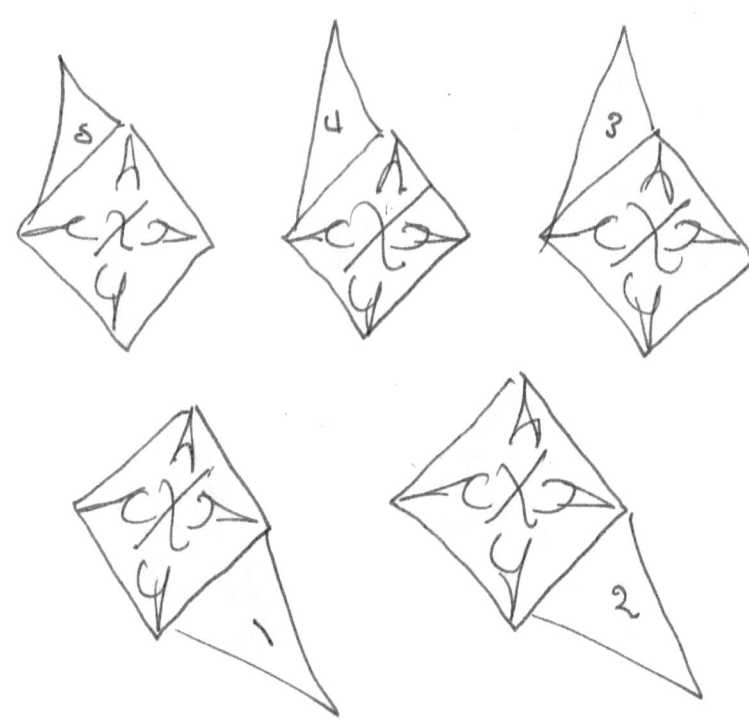

1 2 3 4 5 = Arithmatic and geometric compassess

Arithmatic Compassed = Radio Compasses
Geometric Compasses = Recieving Compasses
→ = Peramid Cone
Expression compass.
[X] = Faculty of Vison
Thus

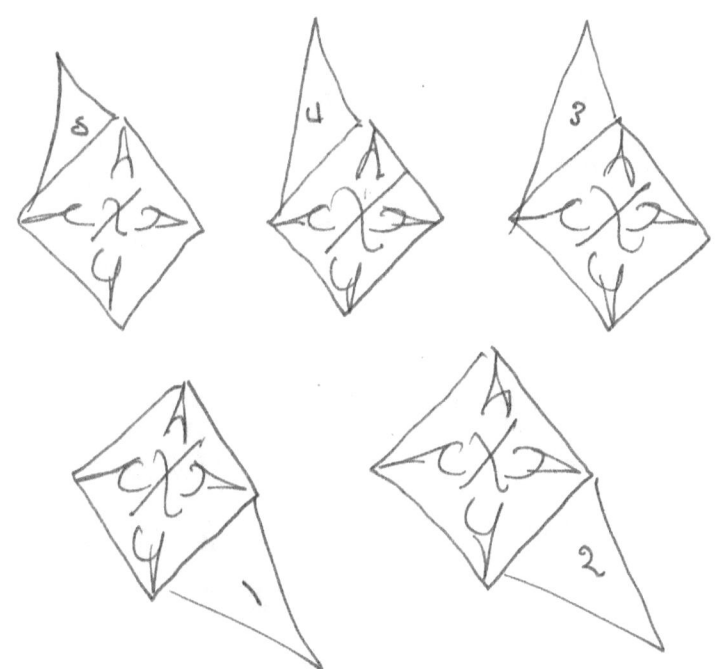

1 2 3 4 5 = Arithmatic and geometric compassess

Arithmatic Compassed = Radio Compasses

Geometric Compasses = Recieving Compasses Expression compass.

→ = Pyramid Cone

[X] = Faculty of Vison

thus/

Thus

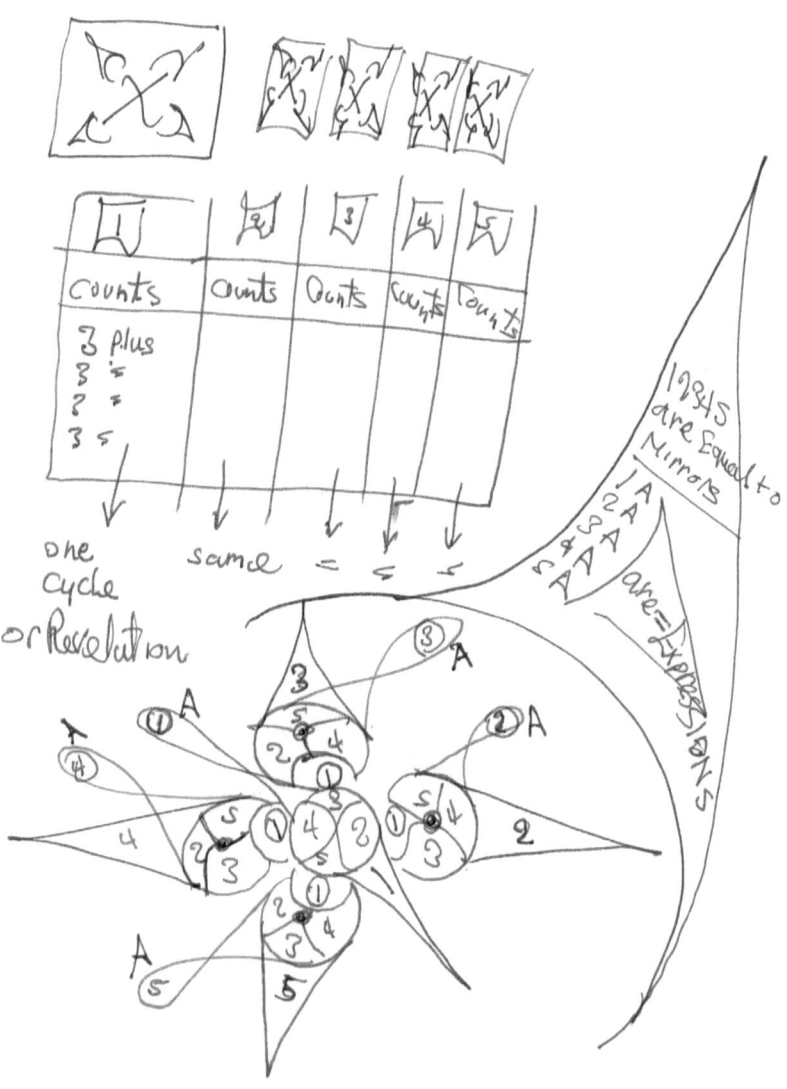

⊞	⊠	③	④	⑤
counts	counts	counts	counts	counts
3 plus				
3 =				
3 =				
3 =				

↓ one cycle or Revolution

↓ same ↓ = ↓ = ↓ =

1,2,3,4,5 are Equal to Mirrors

1A
2A
3A
4A
5A are = Expressions

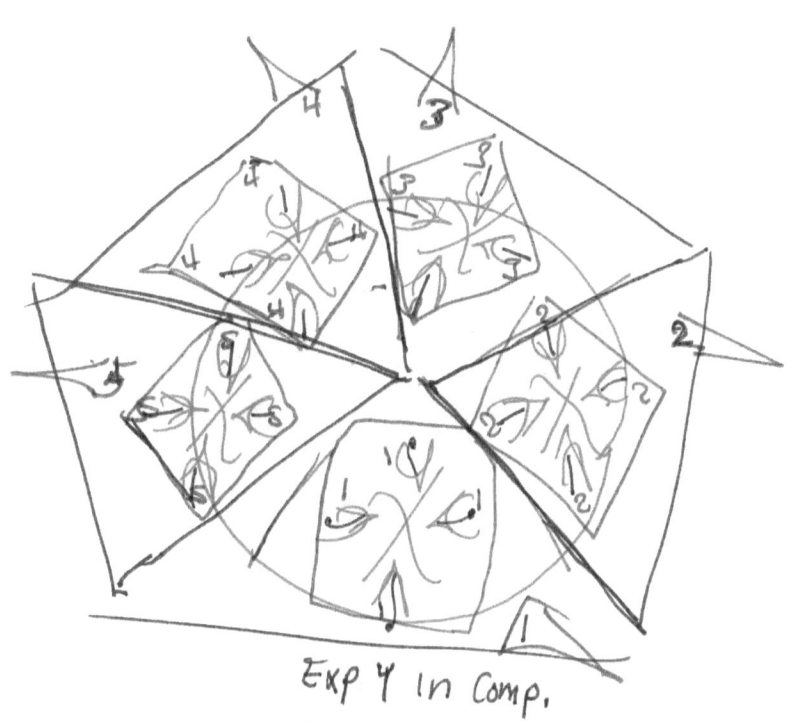

Exp 4 in Comp. 2345

Exp 2 in Compasses 3451
Exp 3 in ——— 4512
Exp 3 in ——— 5123
Exp 4 in ——— 1234
Exp 5 in ——— 1234

← Clock of time.

Equals 1024 squares

fig 19

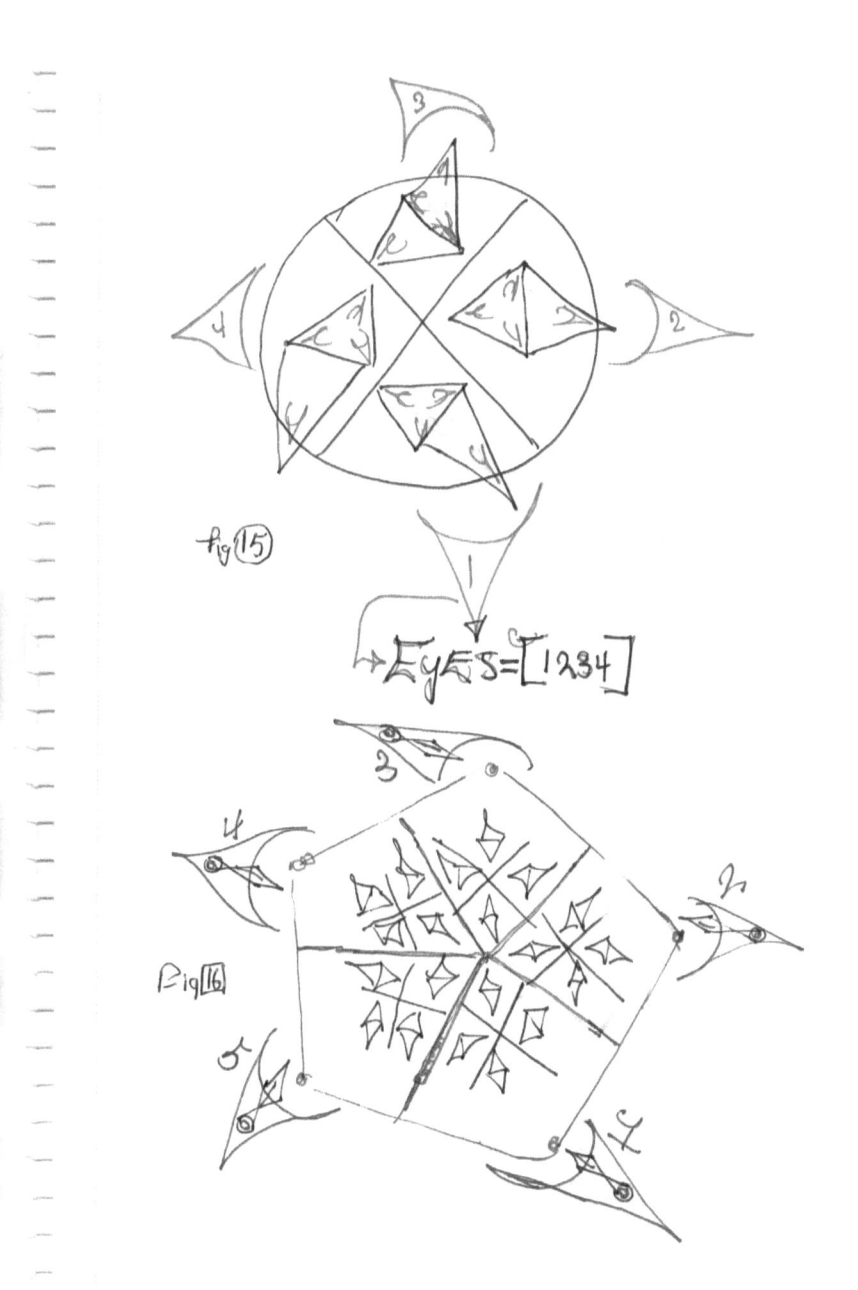

fig(15)

EyES=[1234]

fig(16)

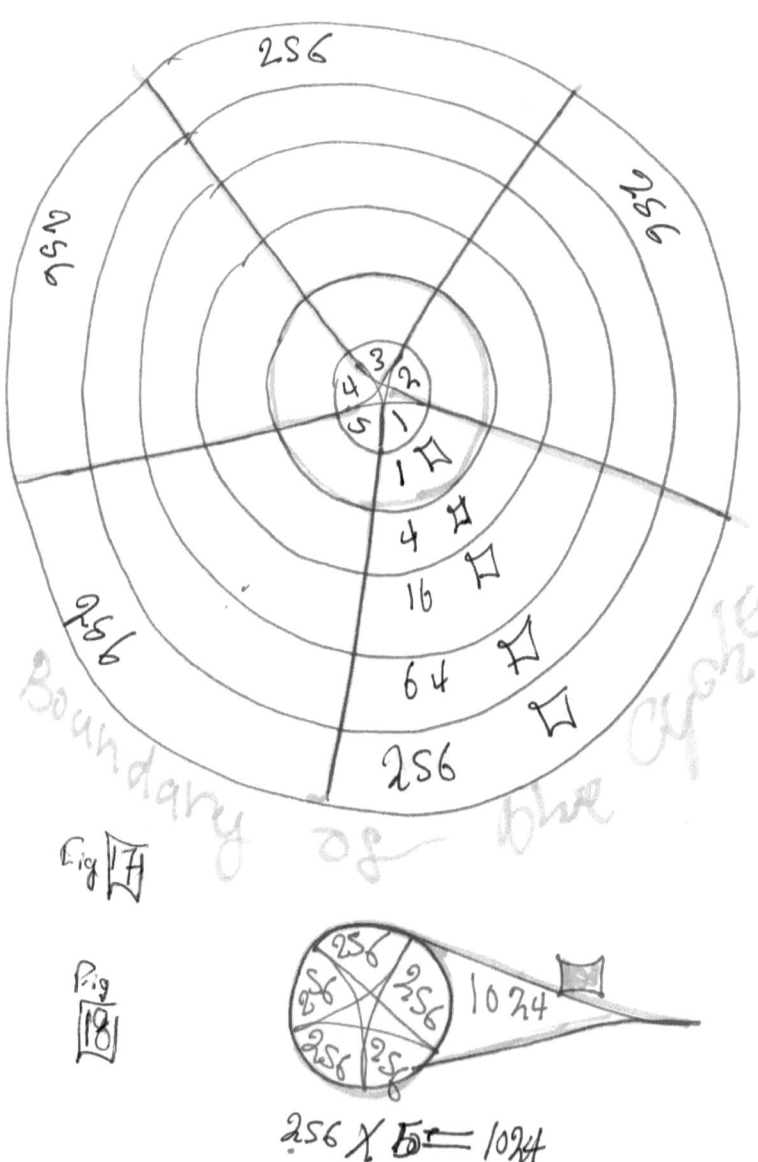

Fig 17

Fig 18

$256 \times 5 = 1024$

APPENDIX

The Text book is an Encyclopaedia of Science and Philosophy; contains a pedagogical innovation that to assist people to use their own mathematics or energy to use their own brains their own pure knowledge which is in Virtue of Possessing It We all Have pure knowledge or human mind contains pure mathematics and that exists as a raw material that requires mathematical methods or instruments to assist individual to apply their own mathematic methods and to use own brains to extract their own inner qualifications and put them into application, creation, construction and opportunity for self-qualification and extraction see the text book without having a text and the guiding instructions to contemplate through the illustrations formed by the text or the text progression in time it is not possible for every ordinary individual to perform mathematical or logical deduction that requires philosophers, only hardworking thinkers can do that analogical performances and they probe to find ways, methods and solutions for the teaching process; the pedagogical question

Thus in this text method the illustrations are equal to means on which individual minds boot on it or tune to it and used as a biometrics and illustrative and translational means

They are for navigation and can be used parallel to the reading text.
Below are Illustration Means

Farek J Sadon

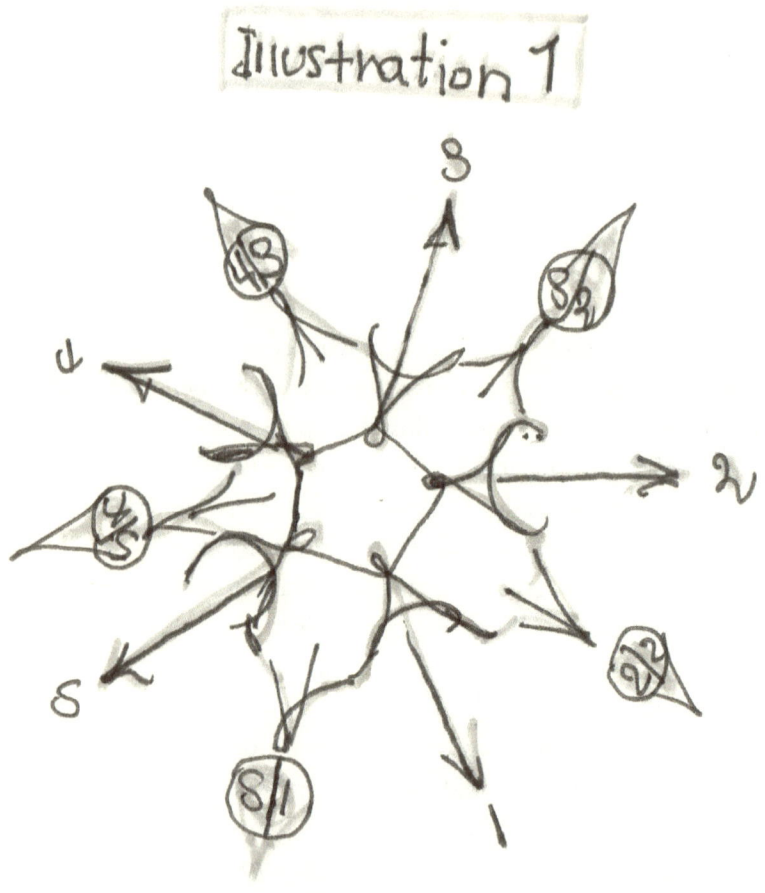

Arrows 12345 are Equal to Pendulum Magnitudes
1.2/2.3/3.4/5.5/5.1 are, Equal to pendulum touches (multiplications) in a Cycle of Time.

Organun

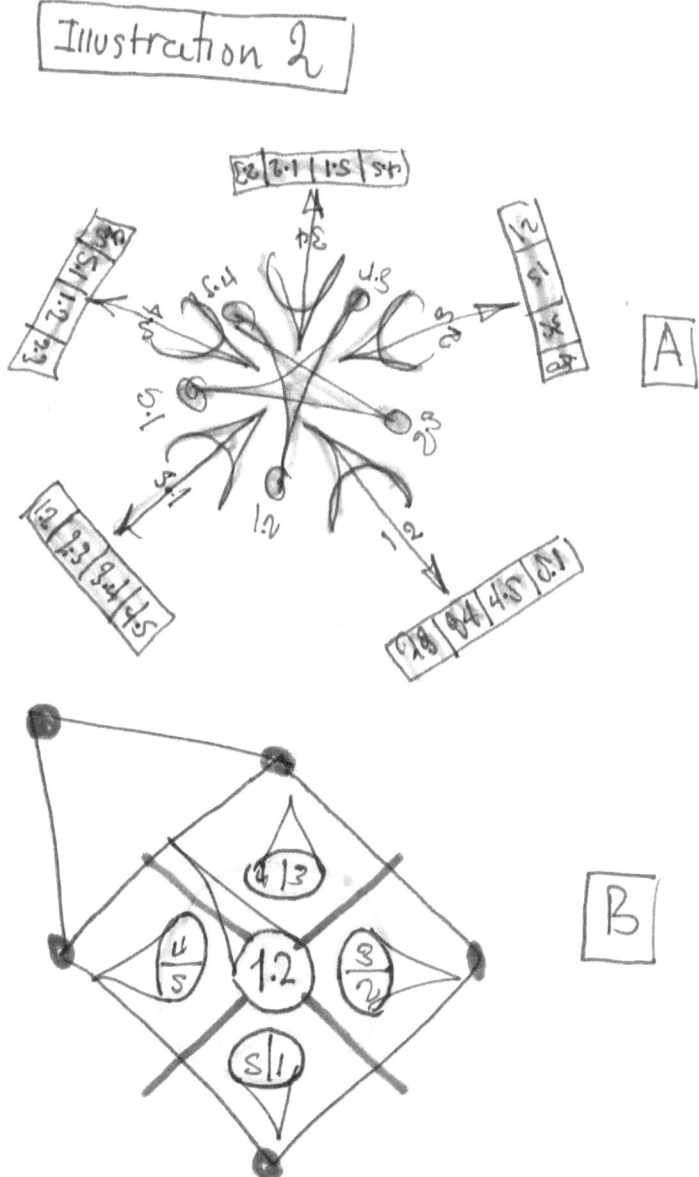

Each Pendulum Motion Acts as Seconds (1.2 Transform to 2.3 transform 3.4 transform 4.5 transform 5.1 then transform 1.2 thus 1.2 completed the Cycle (12: 23,34,45.51) Constructing a Pyramid compass (same for the rest of the 4 Numbers) insert PIC B

Organun Proposition

    d. Everything that Exist equals to structure <u>or</u> composition in time Human apprehension Equals to time in Structure <u>and</u> Composition Apprehension

    E. Everything that exist equals to pendulum *or* Clocks in time Magnet and Magnetic fields is Equal to time in Pendula *and* Clock apprehension

    F. Everything that exist = points or lines in a pentagon Human Grip or apprehension = to time in points and line apprehension

⑧

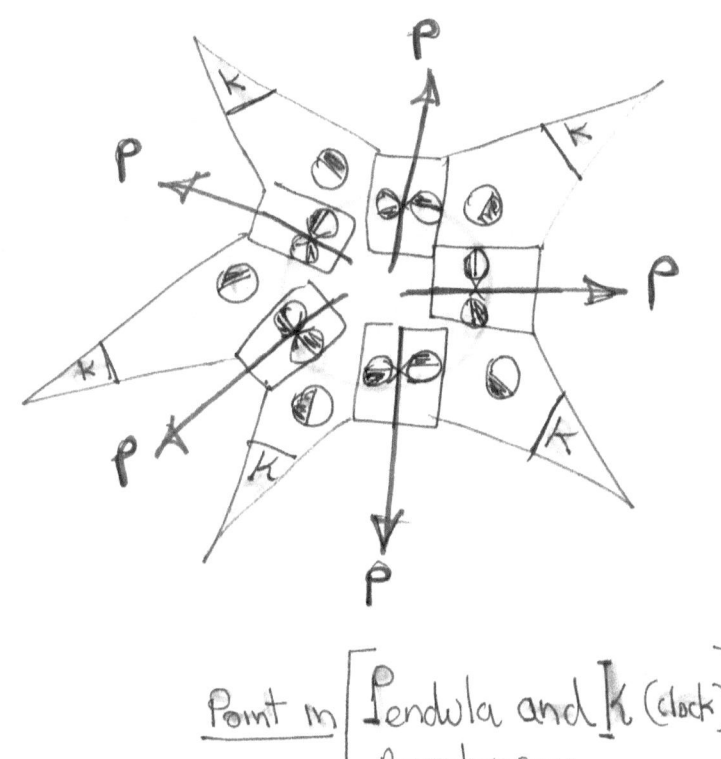

Point in $\begin{bmatrix} \text{Pendula and } \underline{K} \text{ (clock)} \\ \text{Apprehension.} \end{bmatrix}$

④

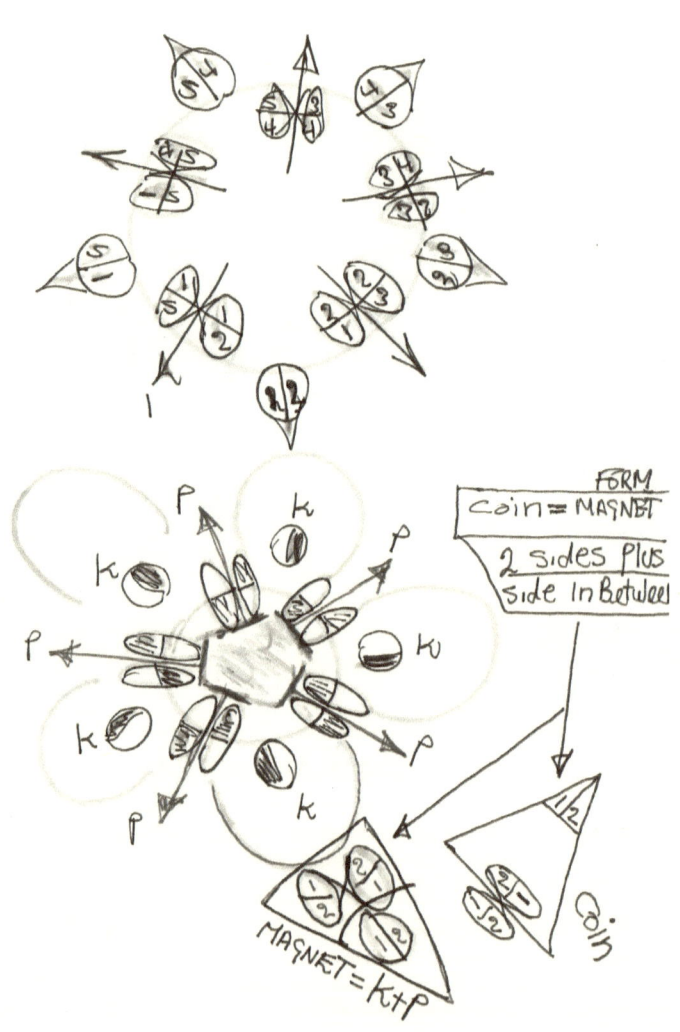

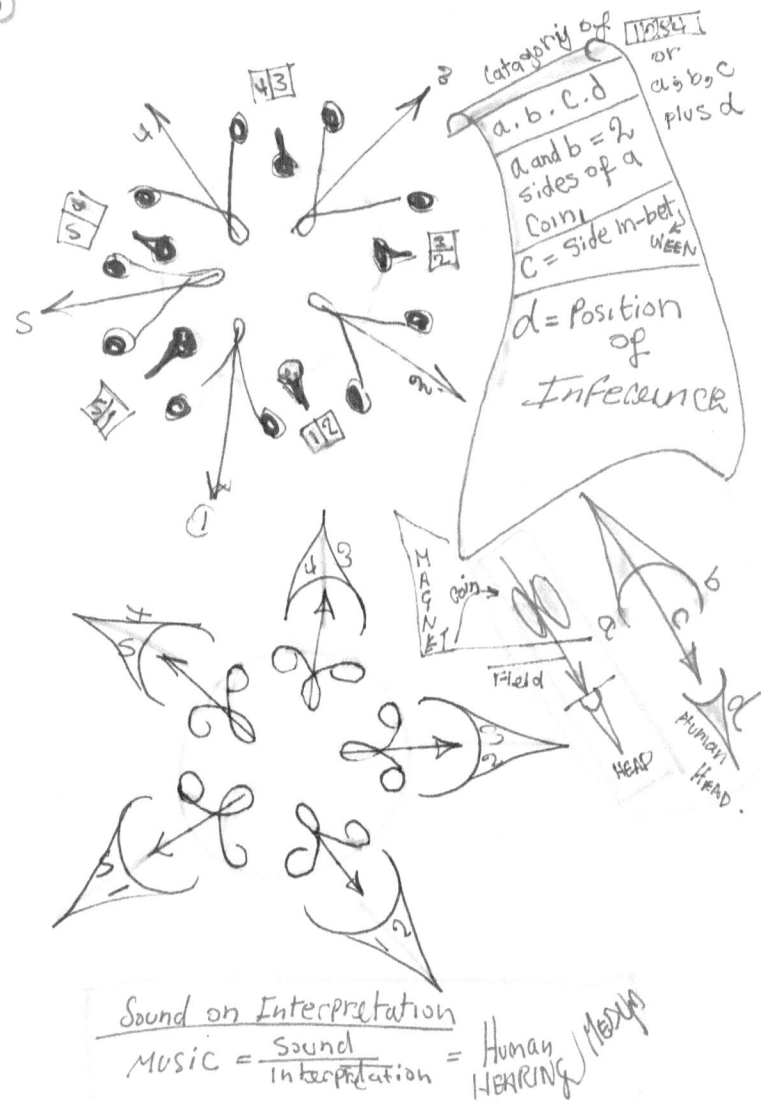

$$\text{MUSIC} = \frac{\text{Sound on Interpretation}}{\text{Sound}} = \text{Human HEARING Medys}$$

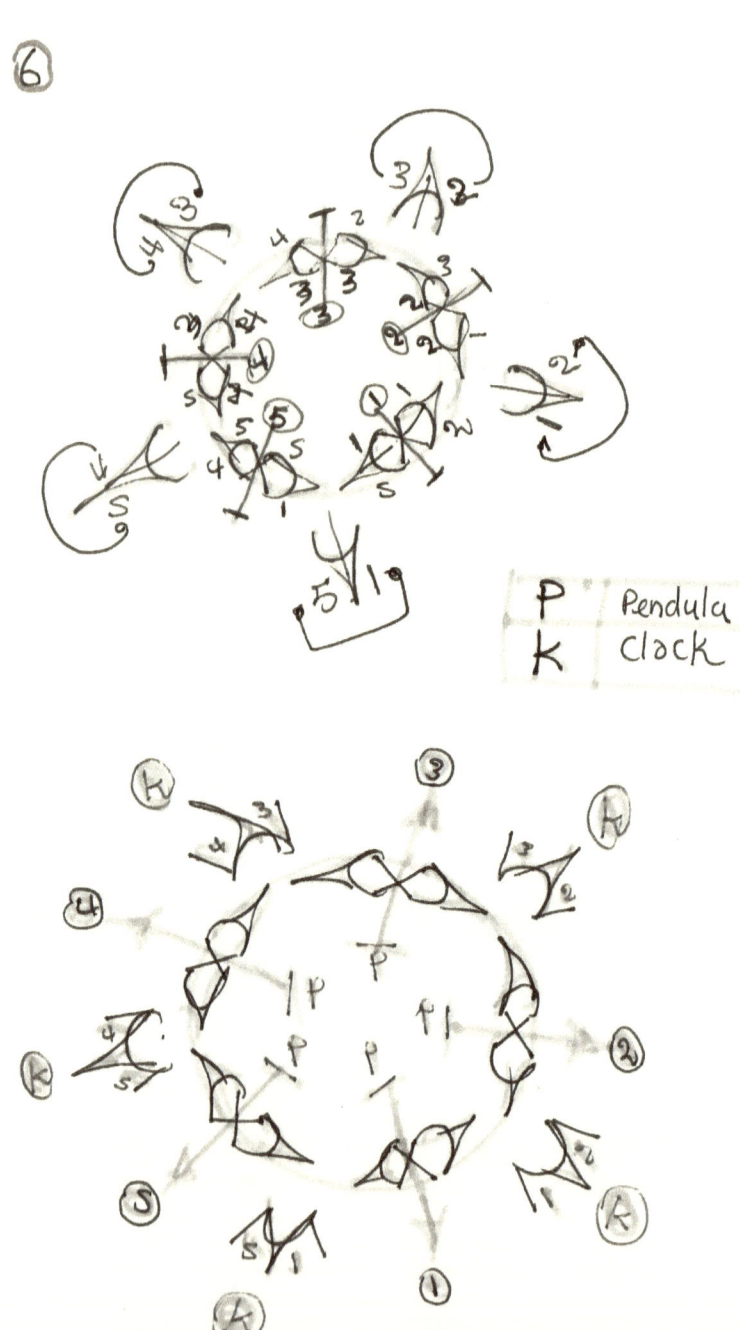

$P = [Puls]s$
$B = [Beats]$

Beat wave and Sound.

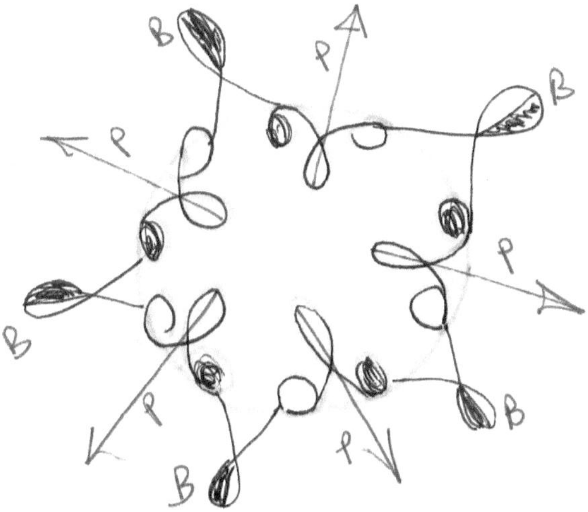

Puls	and	Beats	HEART	BEATS
P		B		

Beat

Puls

Gray
B W

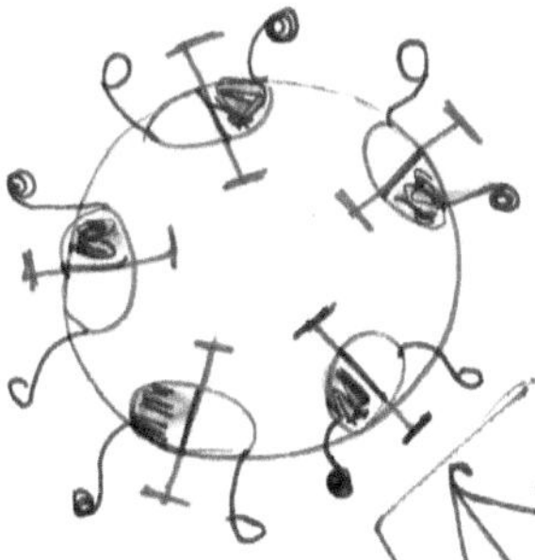

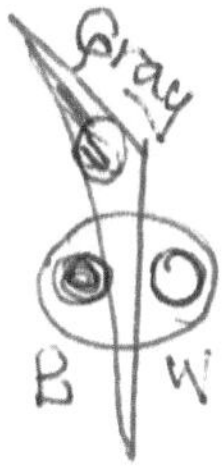

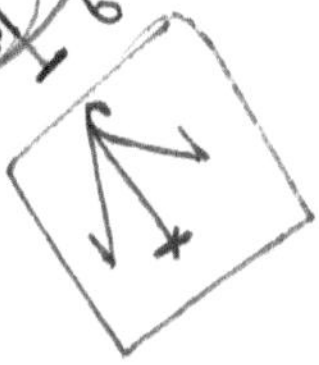

Farek J Sadon

①②③④⑤ are Equal to Pendula
↳ Seconds

①·② ②·③ ③·④ ④·⑤ ⑤·① are = To clocks

K For clocks
P Symbol for Pendula Seconds

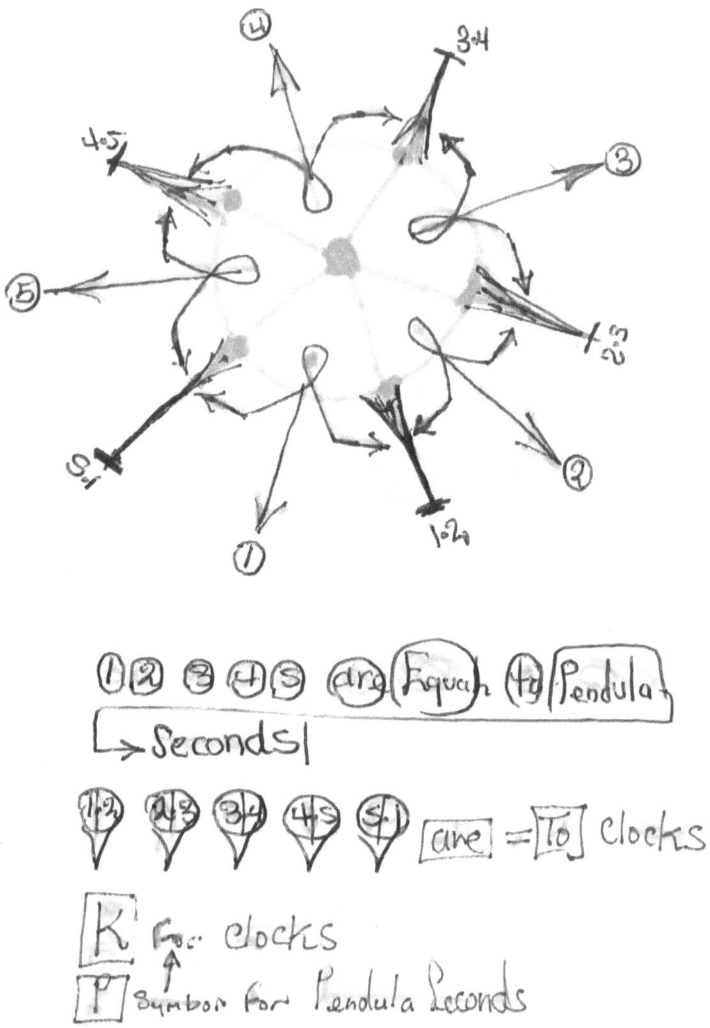

Categories 1,2,3,4
    a. One dimensional
    b. Two dimensional
    c. Three dimensional
    d. Four Dimensional continuum (Time)

Coin Shape equals to (2sides and a line in between) and equals to 3 Dimensional objective reality

And number 4 equates to time ant that of itself equates to Human apprehension

Inferences are organic Axioms that are to differentiate two sides of one coin and to combine both sides together in a form or Shape of a coin simultaneously.

A Diameter Equals to Text or Biometrics of Human Feeling and Equals to units of our Feeling or Consciousness and are Viewed or imagined by Seconds In the vision of time and in fact they are our feelings inside circular transformation and are equals to the strength thus the strength of gravity is our feeling and are the labour that every living creature must perform; thus it is in itself algebra or Syllogistic that progresses from an existing prime provision to the determined end and by necessity an obvious example of this is the obligation of life we all love to live and born without our choice.

1. Pendulum Motion Equal to Its Physical Demonstration (the Strength)
   - Pendulum Motions could be imagined as Bell having a handle in-between two only two polls
   - The Physical Reality of Pendulum Motion is Equal to a piece of Magnet
   - Pendulum motions are represented by lines in Book of Elements and a straight line is an equal distance between two points
   - Thus piece of magnet which is equal to two polls and magnetic field inside the one called Magnetic Field, Radio Active Median, electrodynamics, and mechanics are all terms describing the Magnetic field our Body System Earth and Sky and the four dimensional continuum: *Centre and Sphere and Centre Sphere relation*

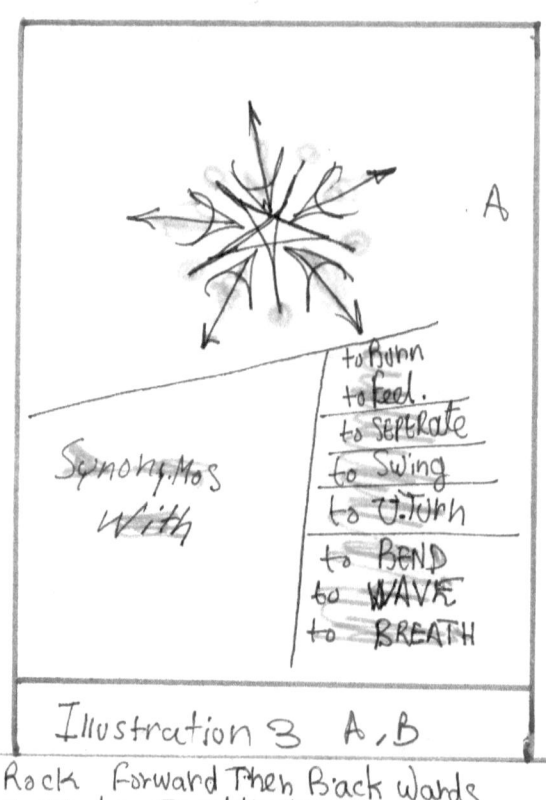

Synonymos
With

to turn
to feel
to seperate
to swing
to U-Turn
to BEND
to WAVE
to BREATH

A

Illustration 3 A, B

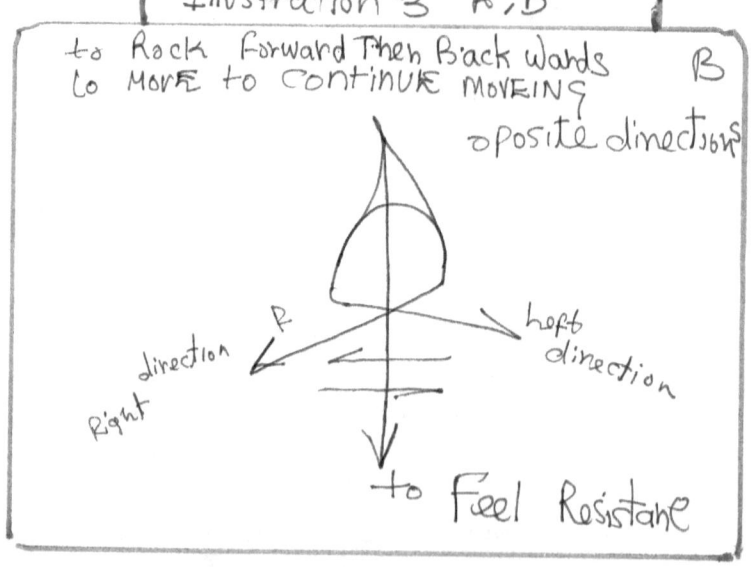

to Rock Forward Then Backwards
to Move to CONTINUE MOVEING
oposite directions

Right direction
Left direction

to Feel Resistane

B

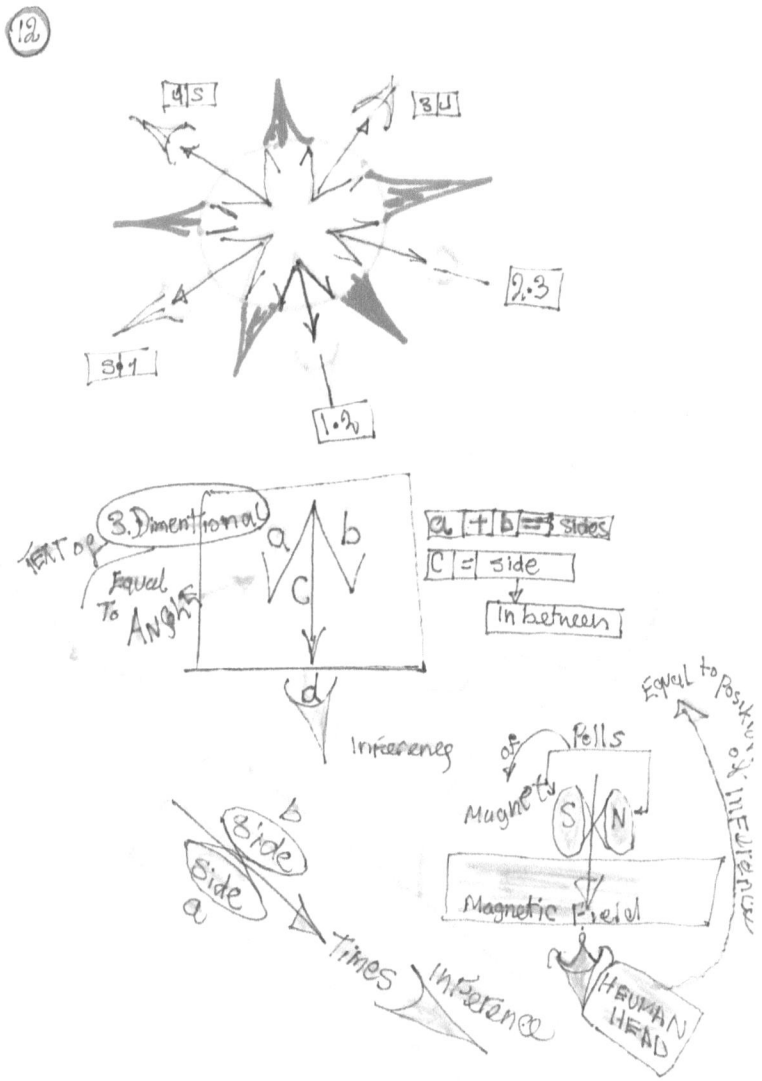

On The Direction of the Current:

the current produced via induction and the light that emits via Force times Resistance process, the light produced inside tube or a law pressure Air container, is Proportional to the magnet, the conductor metal and the human head

Illustration 7

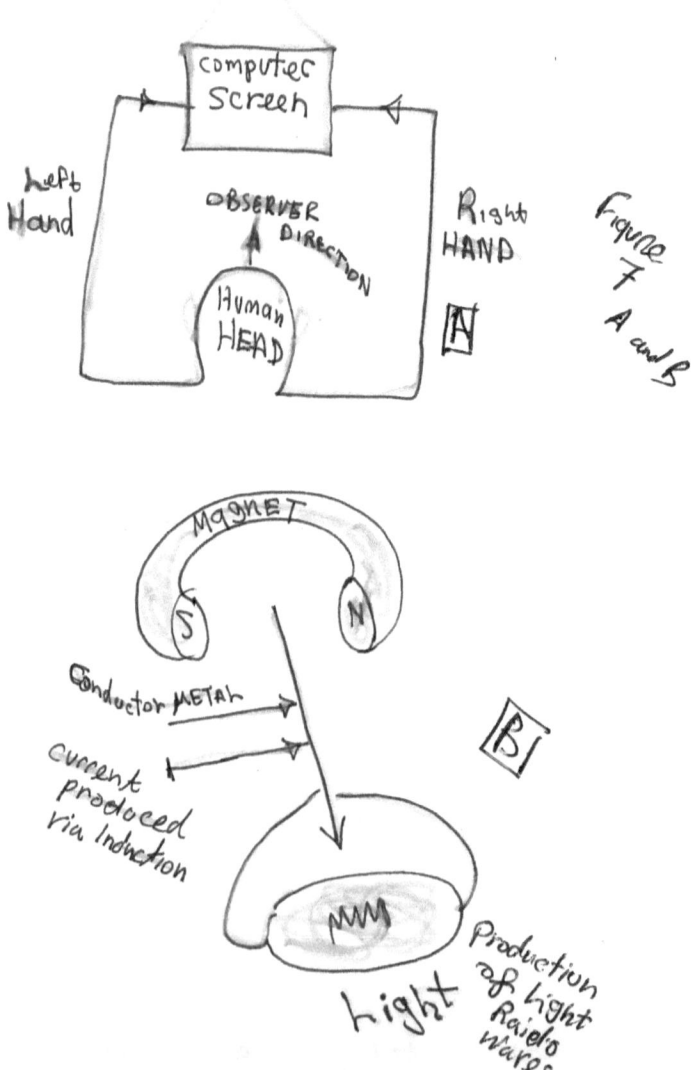

Figure 7 A and B

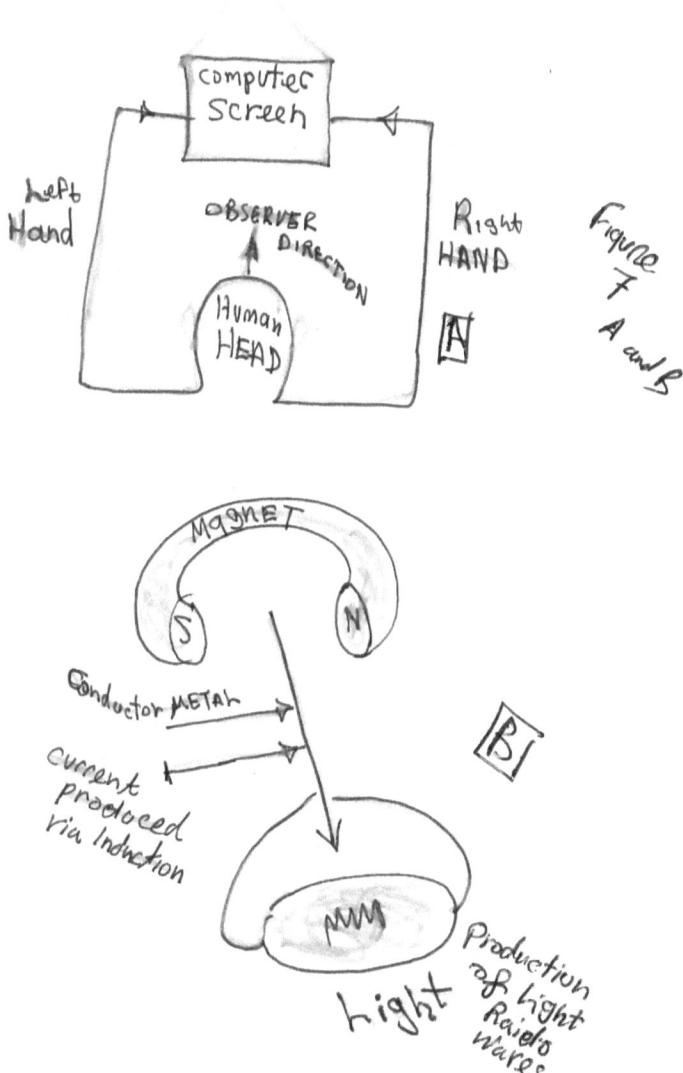

e. Electric Current Produced Via Induction
f. The Nature of Conducting Metals
g. The Law of propagation of Light in Empty Space
h. The Light Produced Via the process of Current Times Resistance (Heater and heated Filament, production of direct current then production of light, the Entity with Dual Nature, Wave particle Dilemma) laser beams, radio waves, microwaves the total application in the domain of communication and observation Technology thus we have three things in the hand as a **Tool of Science**, Magnet, Metallic Conductor, Empty Tube, Optics, Computer TV Screen and a magnet in the hand that can deflect the light produced inside an empty Tube or container and the third line that connecting human mind to the event taking place in the space, outside of Human Body or to the spot of light produced in the CRT or Computer Screen.

Euclidian and nun Euclidian Geometry Dilemma
Postulate Five Dilemmas the 1/2
and 3/2 Ratio

Physical reality of the point and the Function of x

All parallelograms (illustrated in the Illustration Figures) are equal Parallel lines and are Equal Sides of a Magnet and the third arrow line that directed at human hearing media: the head, Back Head and front the Face. **This is an intersection relation, with the current or Electricity produced via induction, Tools used in the process; is [piece of Magnet and current conducting metal] then production of Light, radio activity,** radio waves, TV screens, CRT Cathode Rays and the relation, Electricity and its interaction with human Faculty of vision, the light produced in a vacuum tube is in common notion, relation, interaction with the piece of magnet and our sense of vision

(Light, from the sun or air empty tube) (Piece of magnet via which the electric current, electric charge produced) And human Eyes or Hearing Media the light produced via magnetic media is proportional to human Eye Ears and the piece of magnet and that ratio equal to **3/2**

Illustration 6

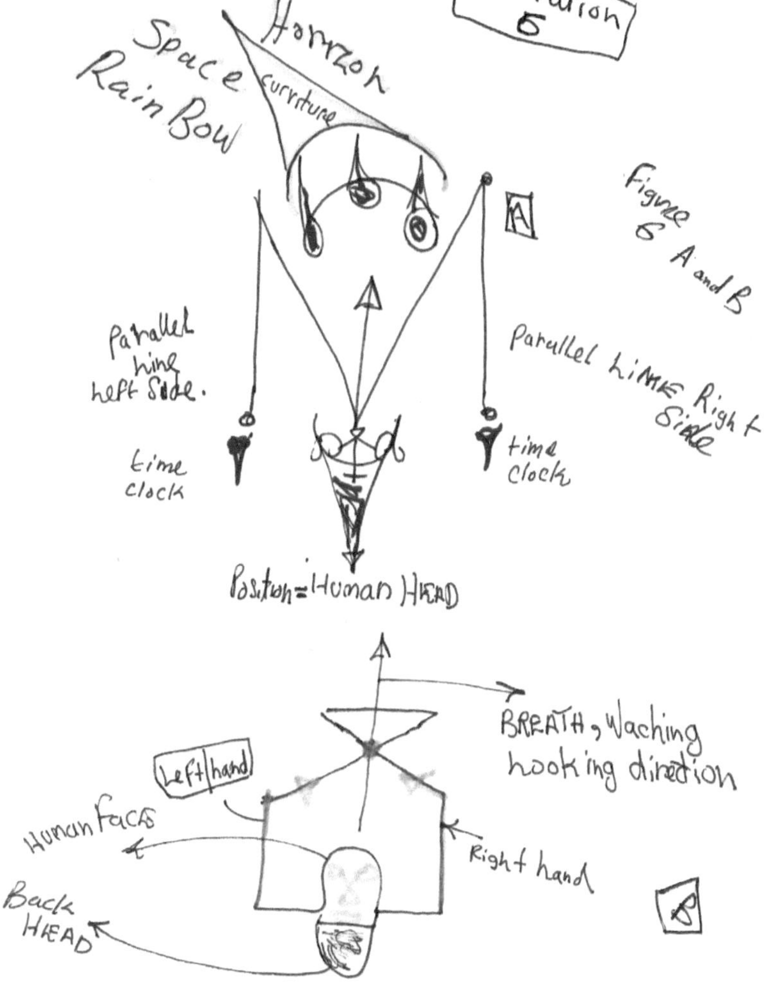

Figure 6 A and B

Space Rain Bow — Horizon, curvature

Parallel line Left Side — time clock

Parallel Line Right Side — time clock

[A]

Position = Human HEAD

BREATH, Waching hooking direction

Left hand

Right hand

Human Face

Back HEAD

[B]

Illustration 5

Figure 5,ABC

SPACE curvature

Curvature of space

Ⓐ

Ⓑ

$\frac{3}{2}$ Ratio

straight line of time

Ⓒ

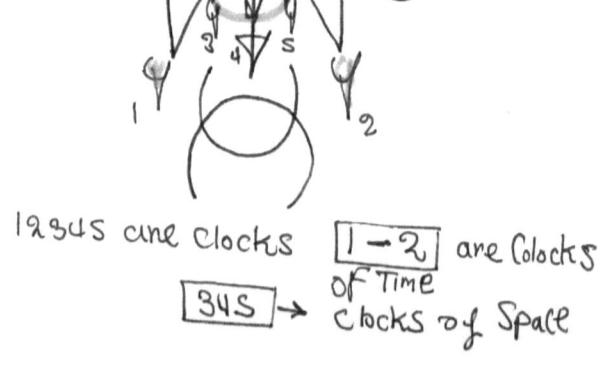

1 2 3 4 5 are clocks

3 4 5 → clocks of Space

1 — 2 are clocks of Time

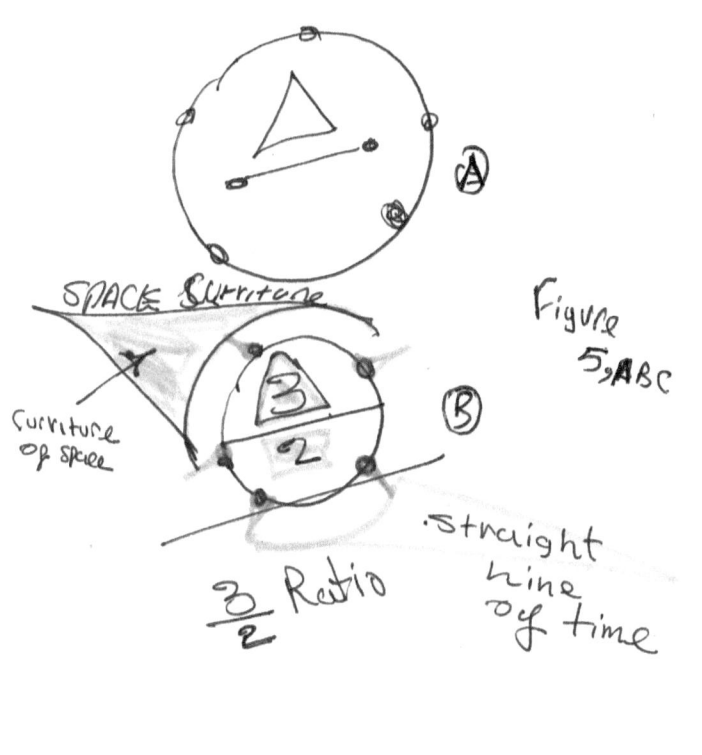

Figure 5, ABC

SPACE curvature

Curviture of space

3/2 Ratio

straight line of time

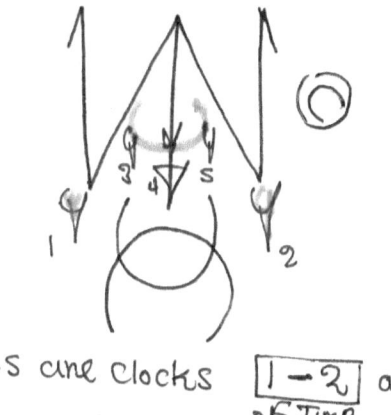

1 2 3 4 5 are clocks

3 4 5 → clocks of Space

1 – 2 are clocks of Time

Illustration 8

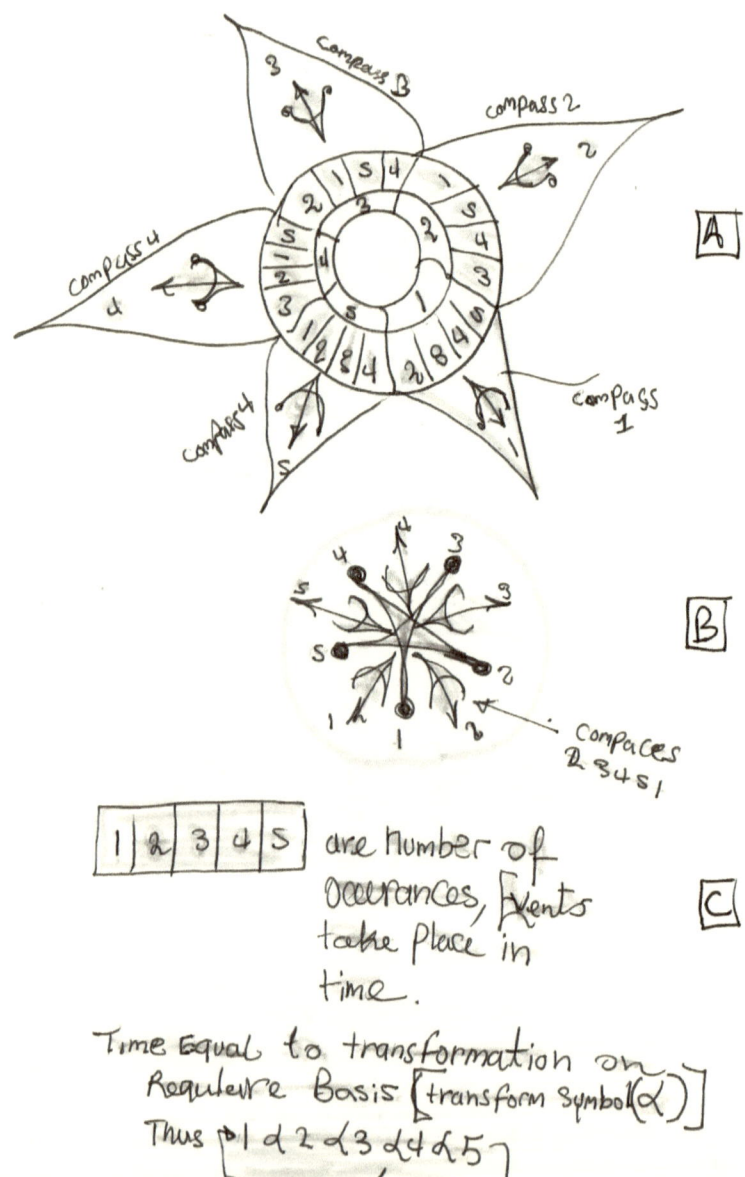

|1|2|3|4|5| are number of occurances, Events take place in time.

Time Equal to transformation on Regulure Basis [transform symbol(α)]

Thus $\boxed{1 \alpha 2 \alpha 3 \alpha 4 \alpha 5}$
         α

Illustration 9

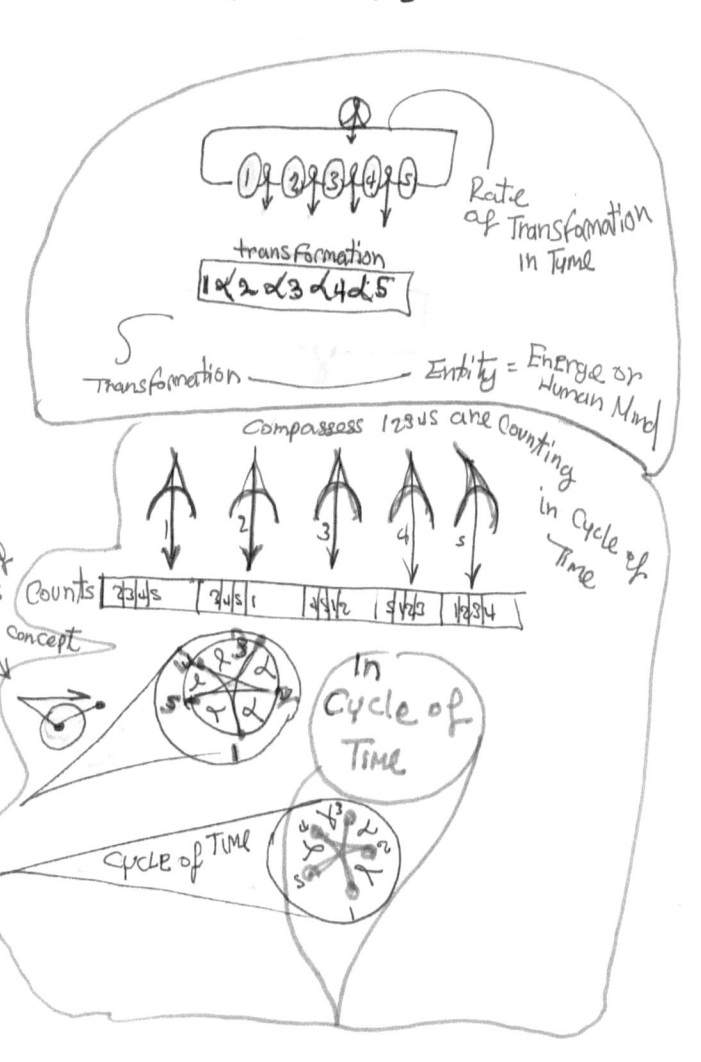

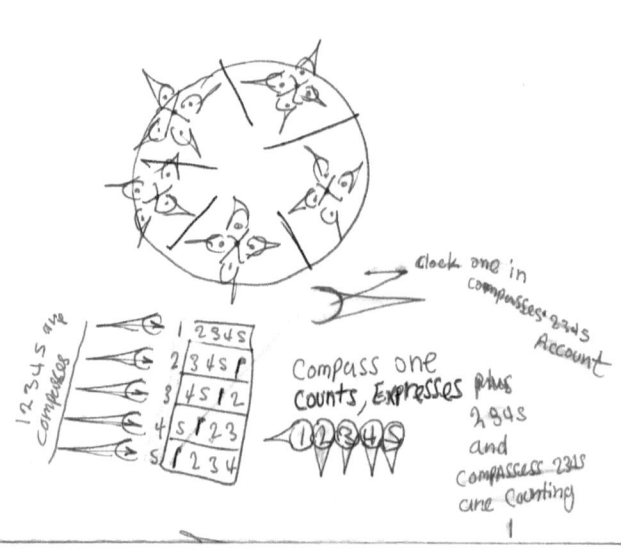

clock one in
compasses 2345
Account

compass one
counts, Expresses plus
2,3,4,5
and
compasses 2345
are counting

1,2,3,4,5 are
compasses

1 2345
2 345 1
3 4512
4 5123
5 1234

1

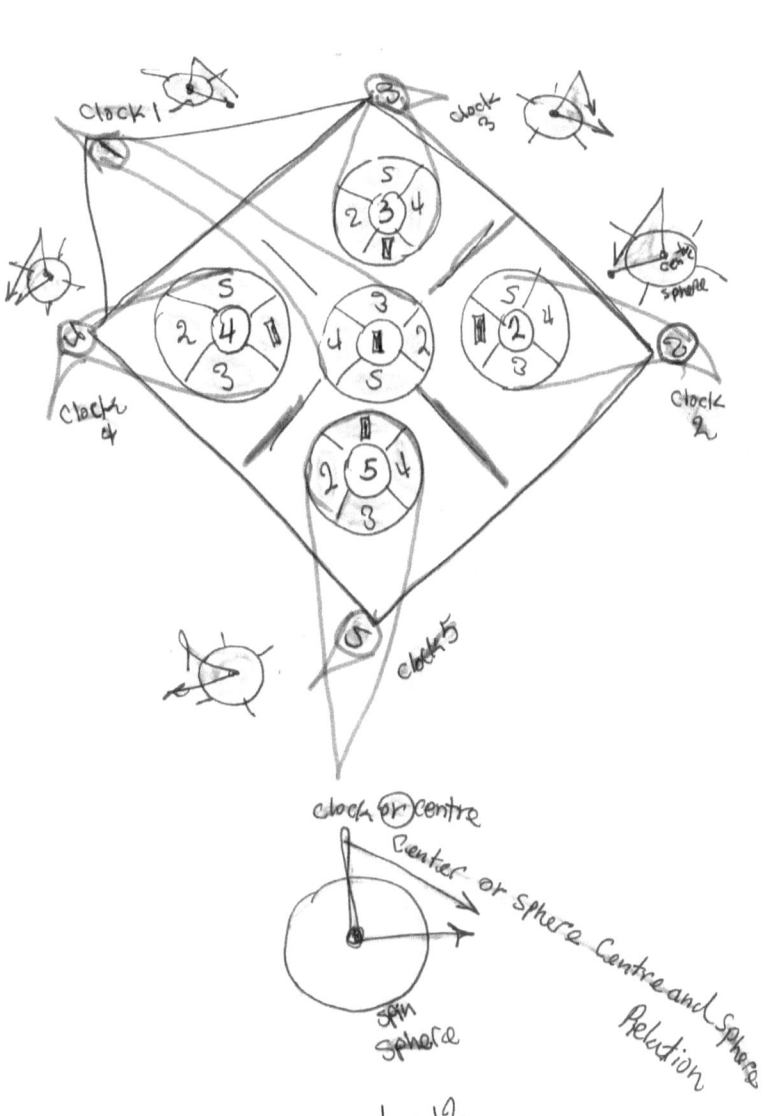

Illustration 12

Illustration 13

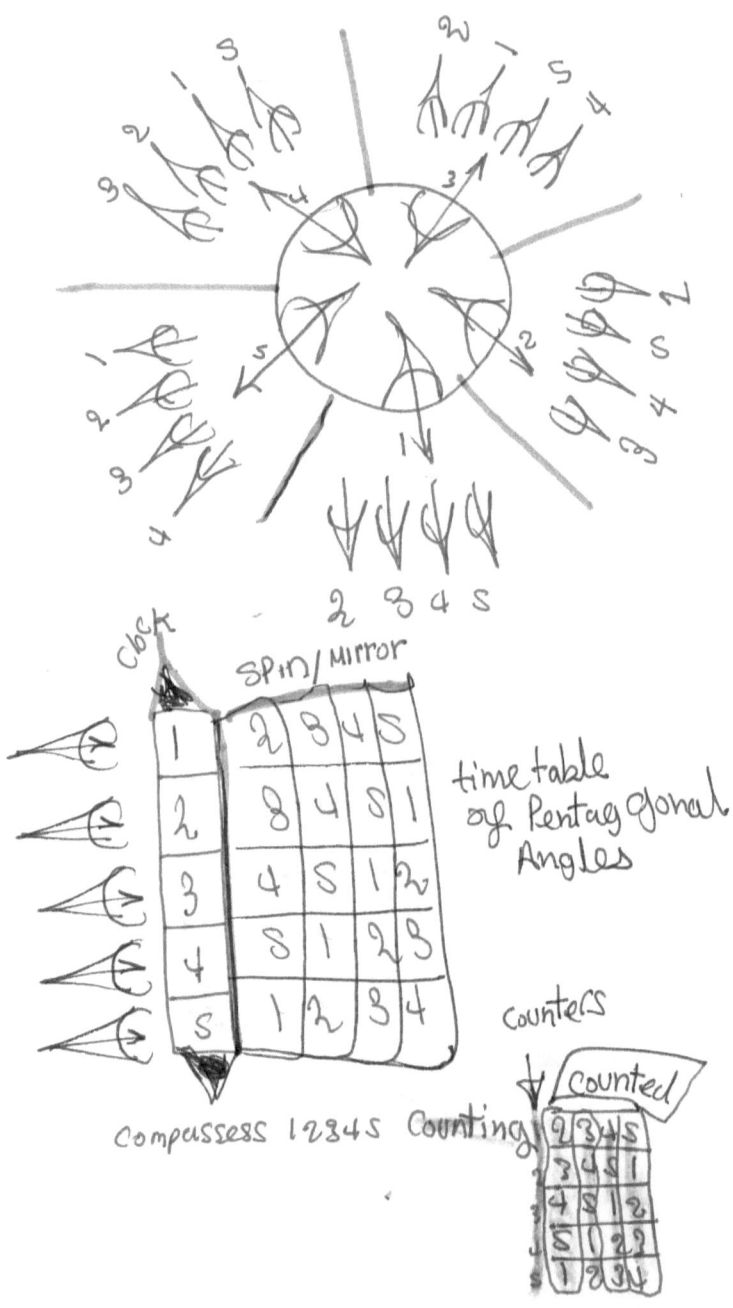

time table of Pentagonal Angles

compassess 12345 Counting counters counted

Organun

The text book concerned with basically two tasks:

3. To pronounce the origin organ of Demonstration and the demonstrative means
4. Answering its possibility and challenge of application.

Demonstration is equal to a Deduction in which the Premises are:
True, Prototype or Primary and Immediate without Middle
See Greek Names, AMESA Medusa with having Hydra Shapes or pentagonal Star.
In the Essene our senses are Mirrors and they are in mirror imaging process.
Our senses are demonstratives and demonstrate our inner feelings
See illustrations Regarding the heart and heart beats
Music is hearing Notes Detected by Human faculty of Hearing

A Pentagon = to a regular polygon with five equal angles
5 times five = 25 and that equal to 25-5=20 thus 20 angles are equal to mirrors
That contain the angles each square (4 Angles are equal to mirror of an angle)
The subtracted five angles are apprehensions that have the twenty other mirror angles
In their apprehension
Thus the pentagon contains five equal squares each contains an angle
Square numbers are = to 4 angles
$4*4=16-4 = 12$ in the apprehension of the subtracted four angles
4 square times 5 equal to 60 in the apprehension of 20
Thus 60 exist by virtue of the five real angles and the 20 mirror angles and the sixty angles contain in the five squares each with four triangles; $12/4 = 3$ x4 =12
Each triangle has 3 numbers; $3x3= 9-3= 6/3 = 2x3$ = hexagonal = $24/4 =6$ x4 =24 x5 120/ 5 = 24 in five squares and 20 triangles and that al equals to plain of human mind
In a semi-sphere

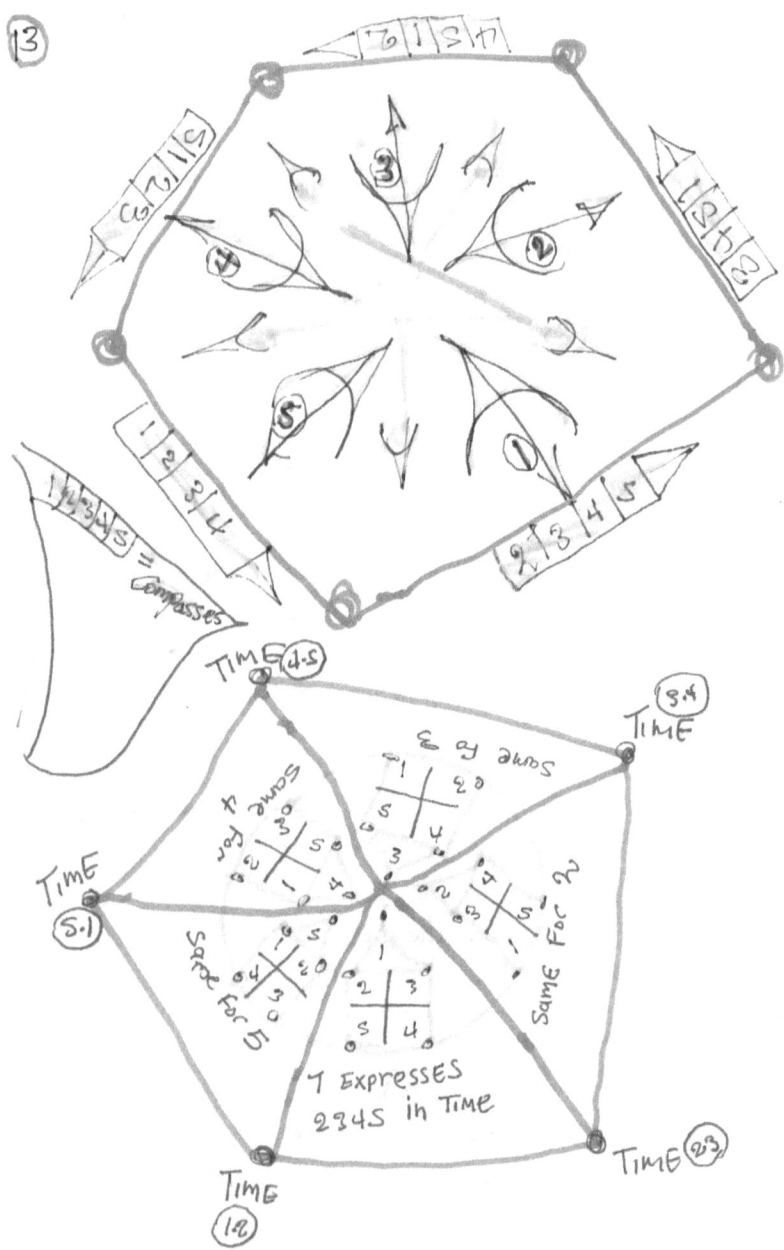

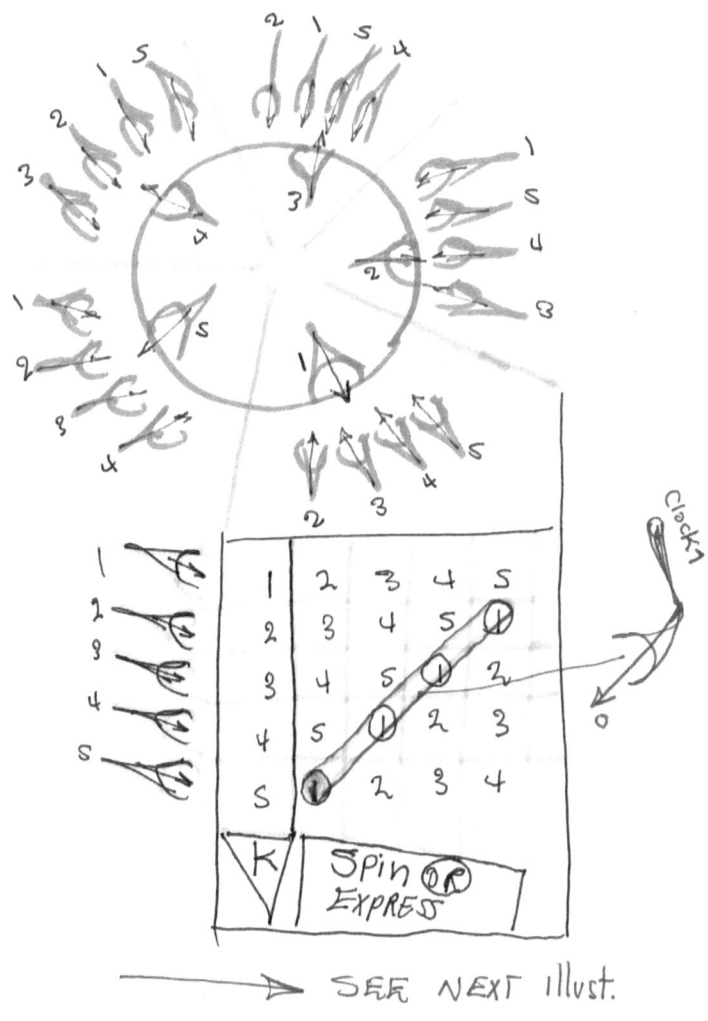

SEE NEXT illust.

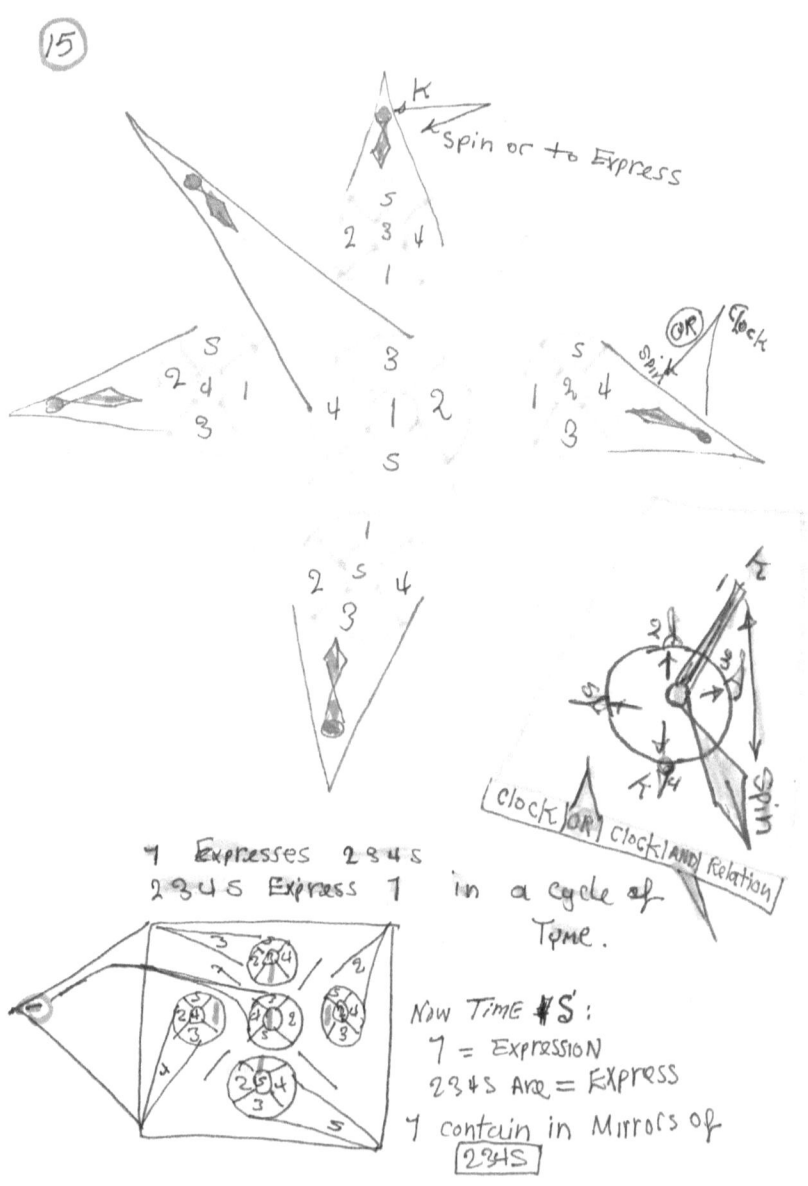

(16)

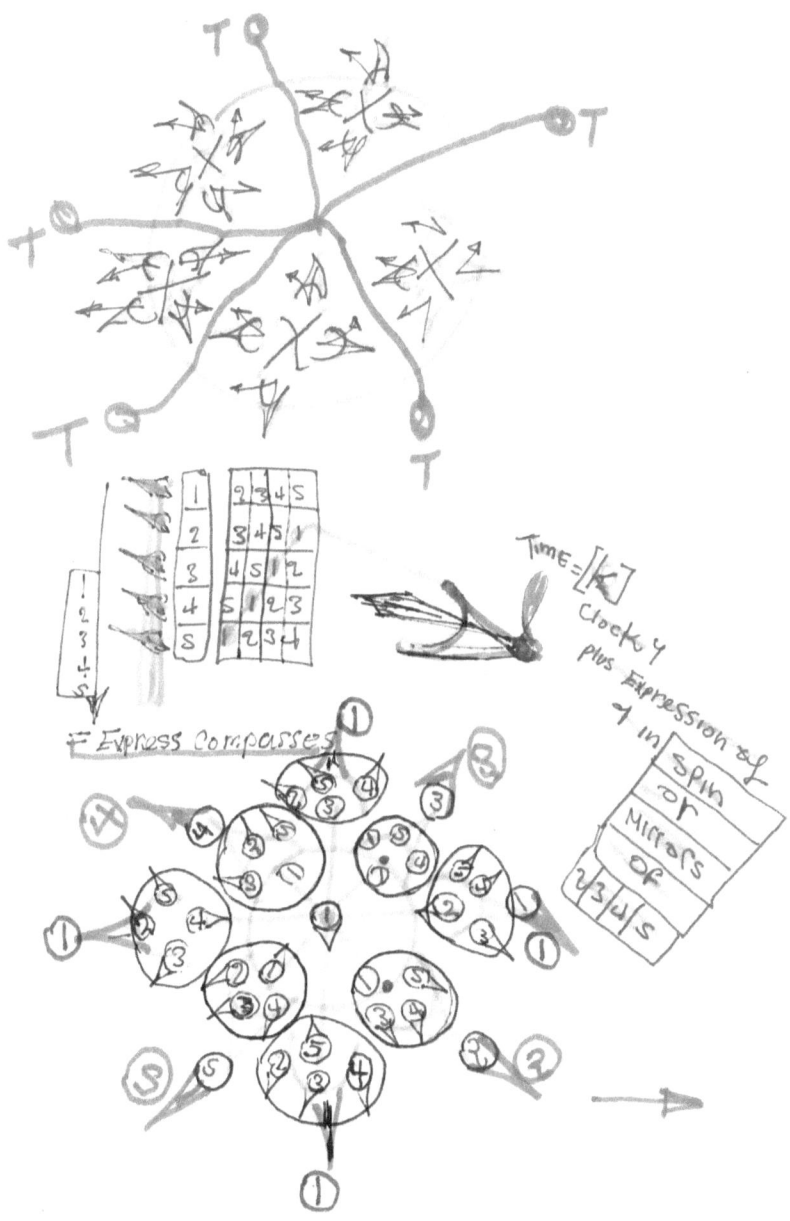

F Express Compasses

Time = [k]
Clocks 4
plus Expression
4 in

| Spin |
| or |
| Mirrors |
| of |
| 2 | 3 | 4 | 5 |

⑰

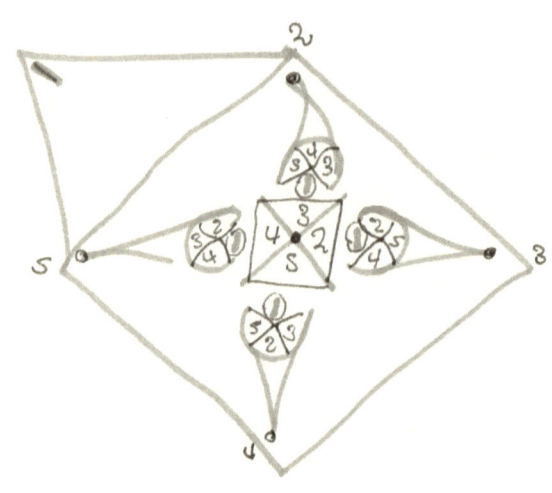

$$\begin{bmatrix} 1 \text{ expresses } 2345 \\ 2345 \text{ Express } 1 \end{bmatrix} \text{In Time}$$

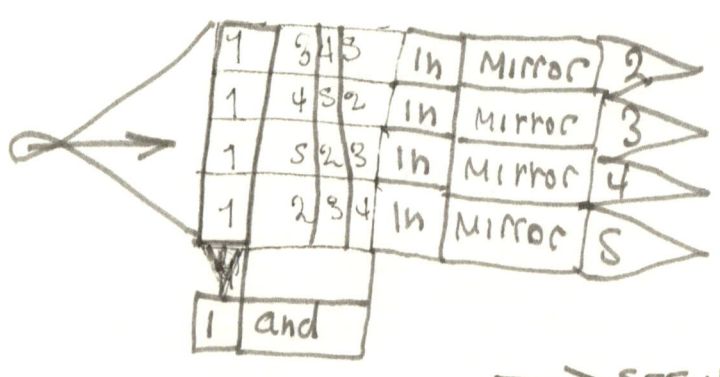

1 and

→ SEE NEXT

⑱

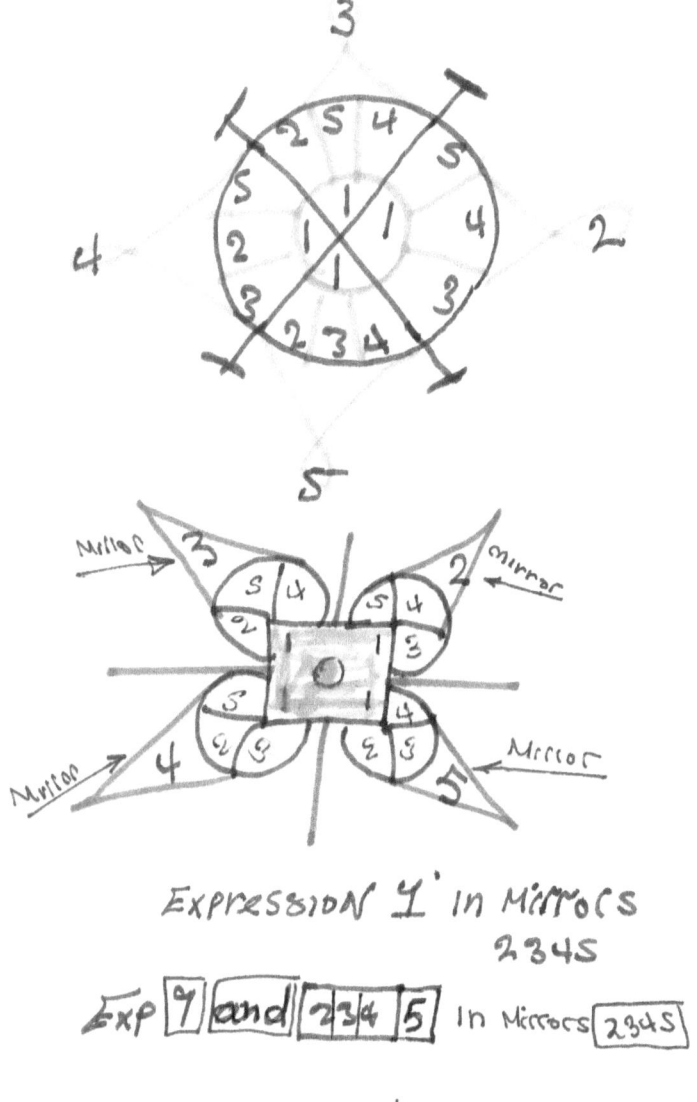

EXPRESSION 1 IN MIRRORS 2345

Exp |1| |and| |234| |5| IN MIRRORS |2345|

to ⟶ Next

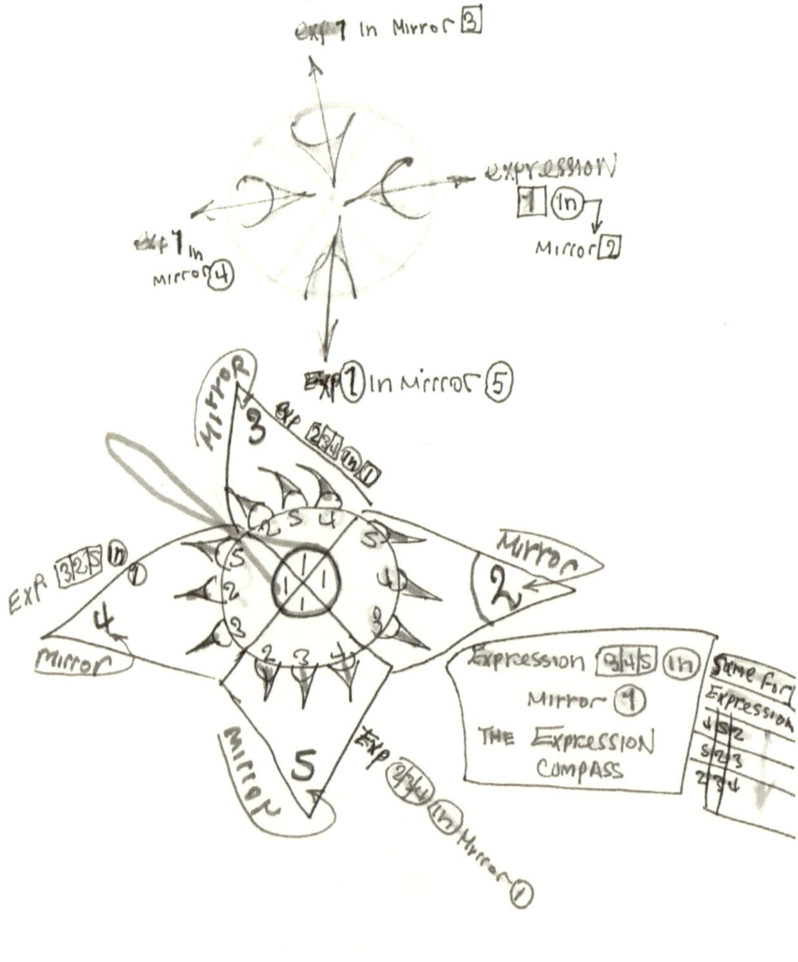

Exp 1 In Mirror 3

expression 1 In → Mirror 2

exp 1 In Mirror 4

Exp 1 In Mirror 5

Mirror 3

EXP 3,2,5 In 1

Mirror 4

Mirror 2

Mirror 5

Exp 2,4,5 In Mirror 1

Expression 3,4,5 In Mirror 1

THE EXPRESSION COMPASS

Same for	
Expression	
1	2
5	3
2	4

Expression COMPASS [7]
counts [12] numbers
in cycle of TIME = one
Minitue

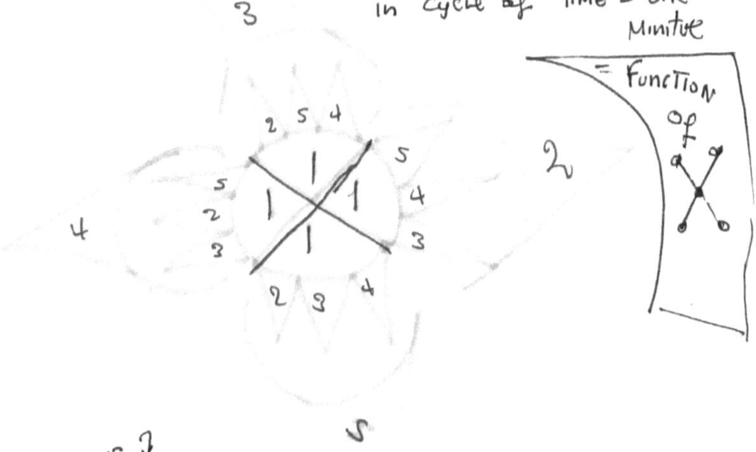

= Function of

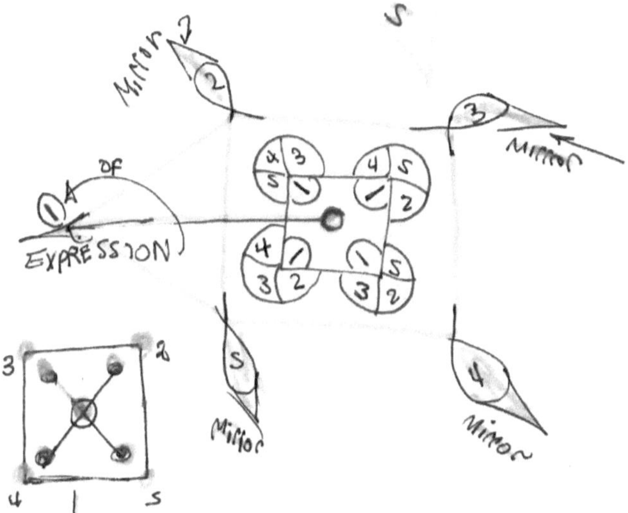

Expression [7] in Mirrors of [2][3][4][5]

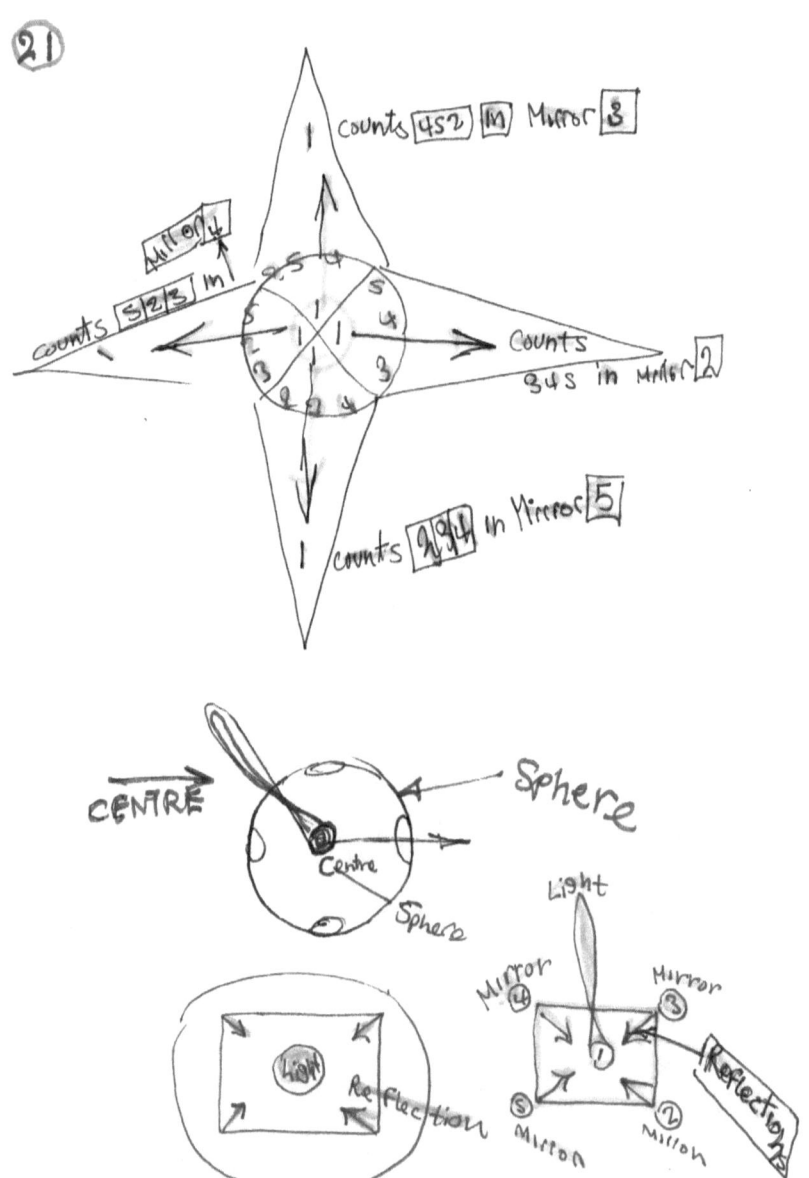

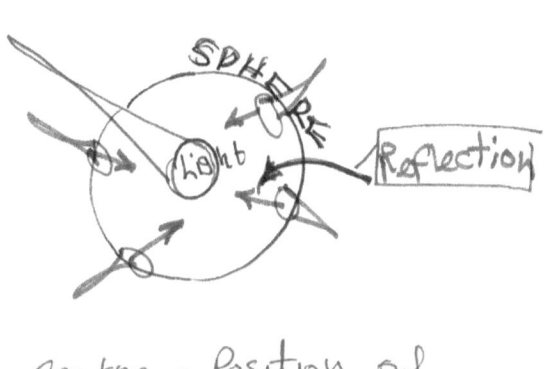

Centre = Position of

| Light | Sound |

SPHERE = Position of Reflection and Equals To Expression of Light or Sound Entity

THUS central = Image
↓
That Contained in SPHER and SPHERE is Equal to its Mirror.

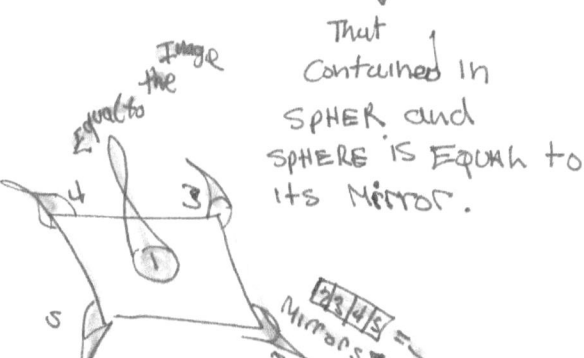

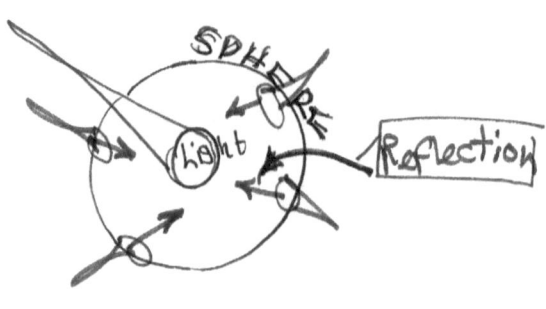

Centre = Position of
| Light | Sound |

SPHERE = Position of Reflextion and Equals To Expression of Light or Sound Entity

THUS Central = Image
↓
That Contained in SPHER and SPHERE is Equal to its Mirror.

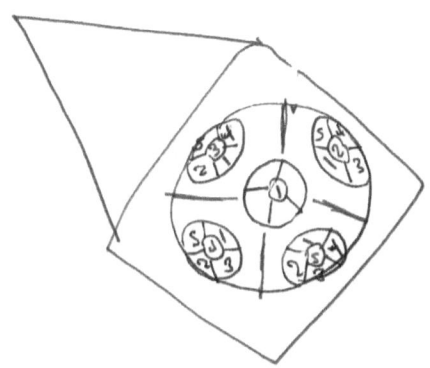

Time is 1 = Focus and
[2|3|4|5 | are] = to Eye or
↳ = Mirror

← [1|2|3|4|5] = sequence of Focah Points
in (Time) in Arethmetic and
   Geometric compasses
↳ of Number one [1]

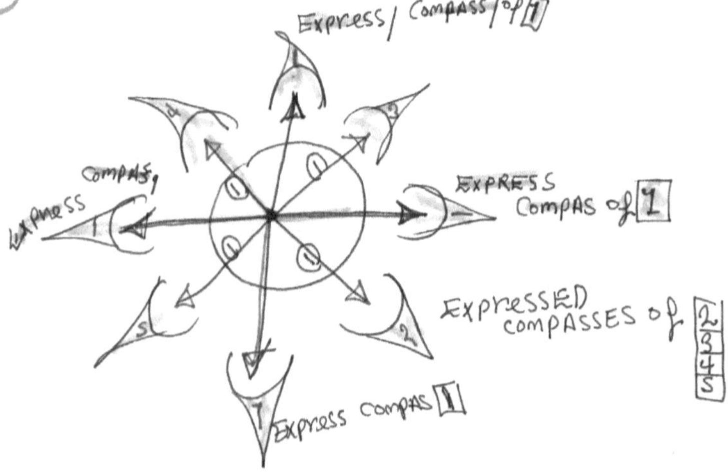

25

Express/Compass/of [1]

Express Compass [1]

Express Compas of [1]

Expressed Compasses of [2,3,4,5]

Express Compas [1]

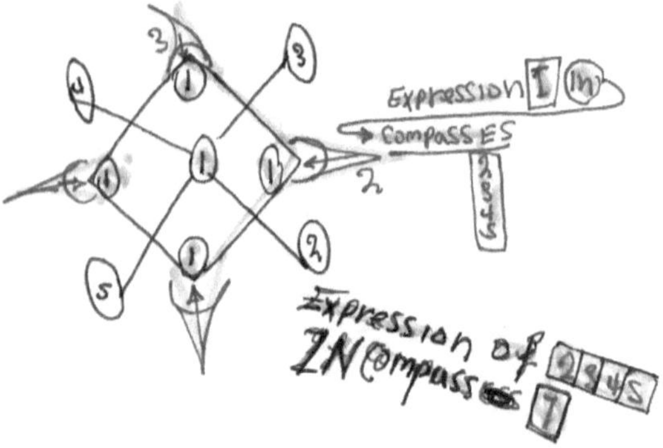

Expression [1] (m)
→ Compasses

Expression of IN Compasses [2,3,4,5] [1]

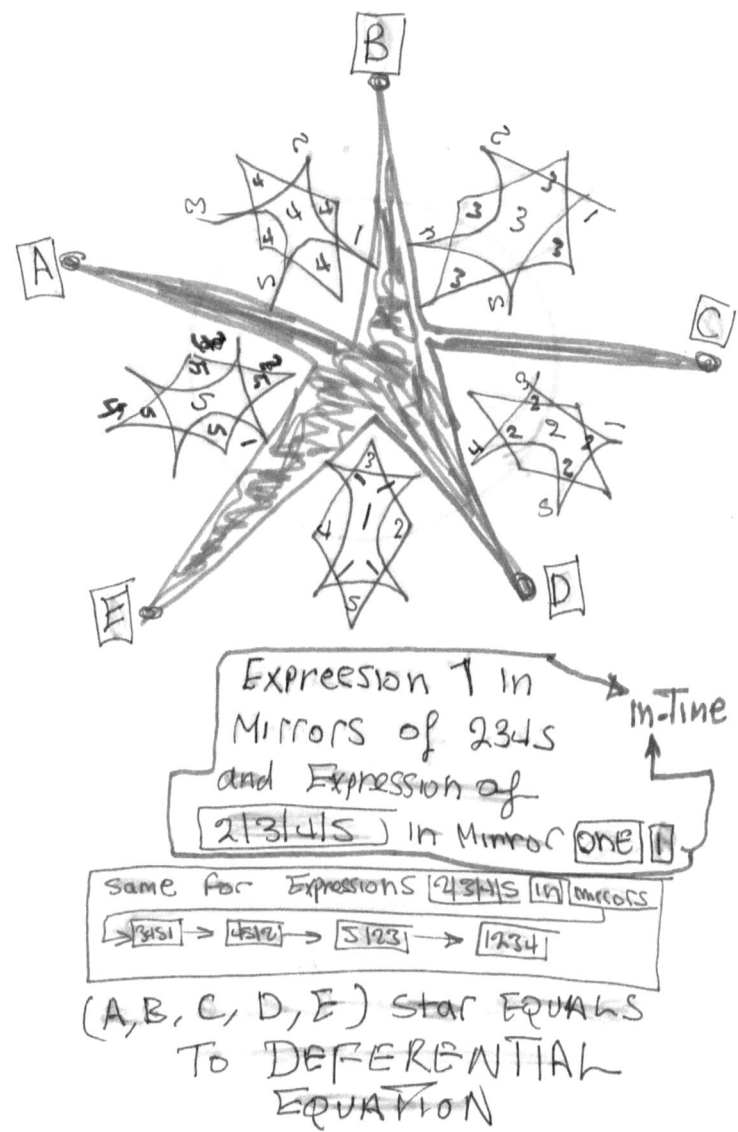

Expreesion 1 in Mirrors of 2345 and Expression of 2|3|4|5 in Mirror ONE

Same for Expressions 2345 in mirrors
3451 → 4512 → 5123 → 1234

(A, B, C, D, E) Star EQUALS TO DEFERENTIAL EQUATION

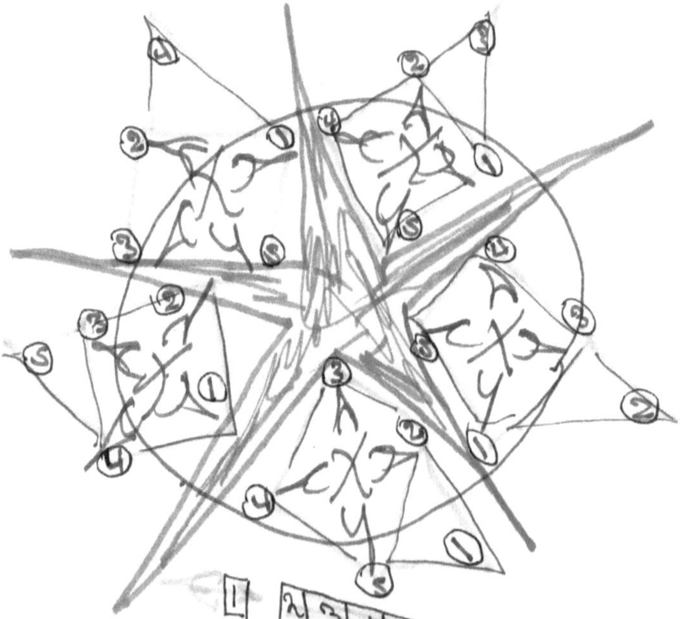

1	2	3	4	5
2	3	4	5	1
3	4	5	1	2
4	5	1	2	3
5	1	2	3	4

TIME IS clock 1 of compasses one 1
in
in spin of compasses 2345

Clock 1 in
Compasses 2 3 4 5

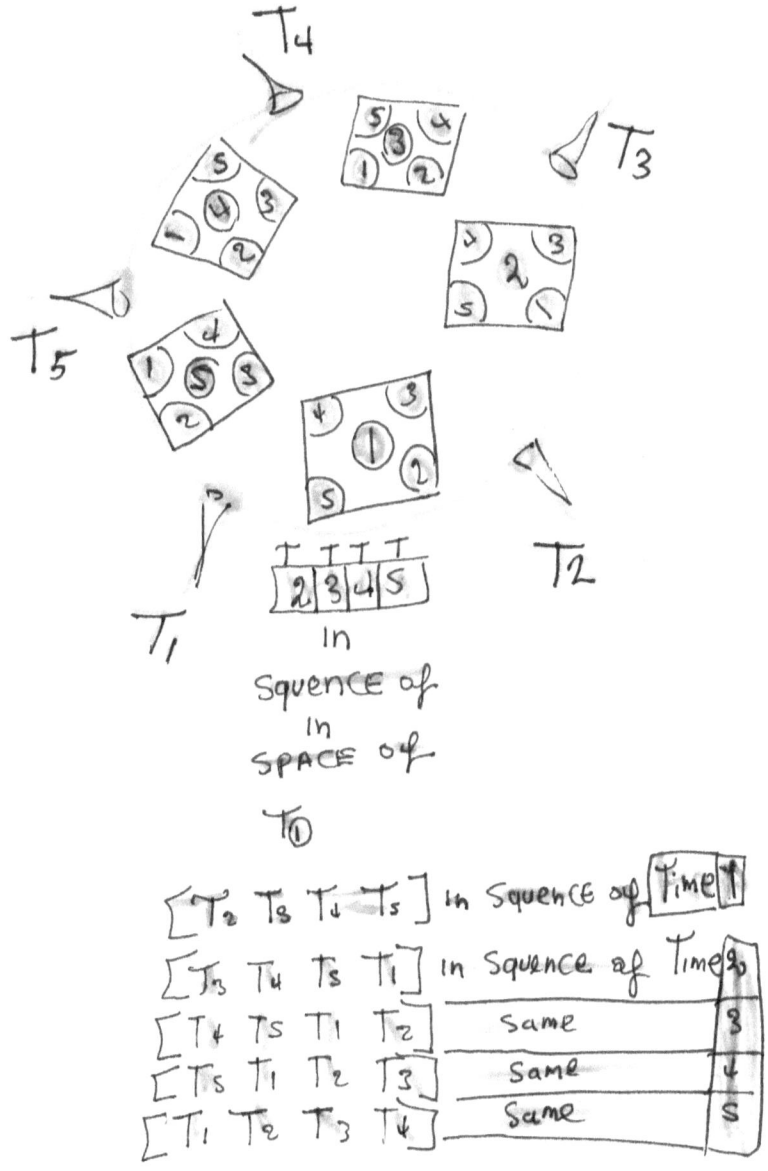

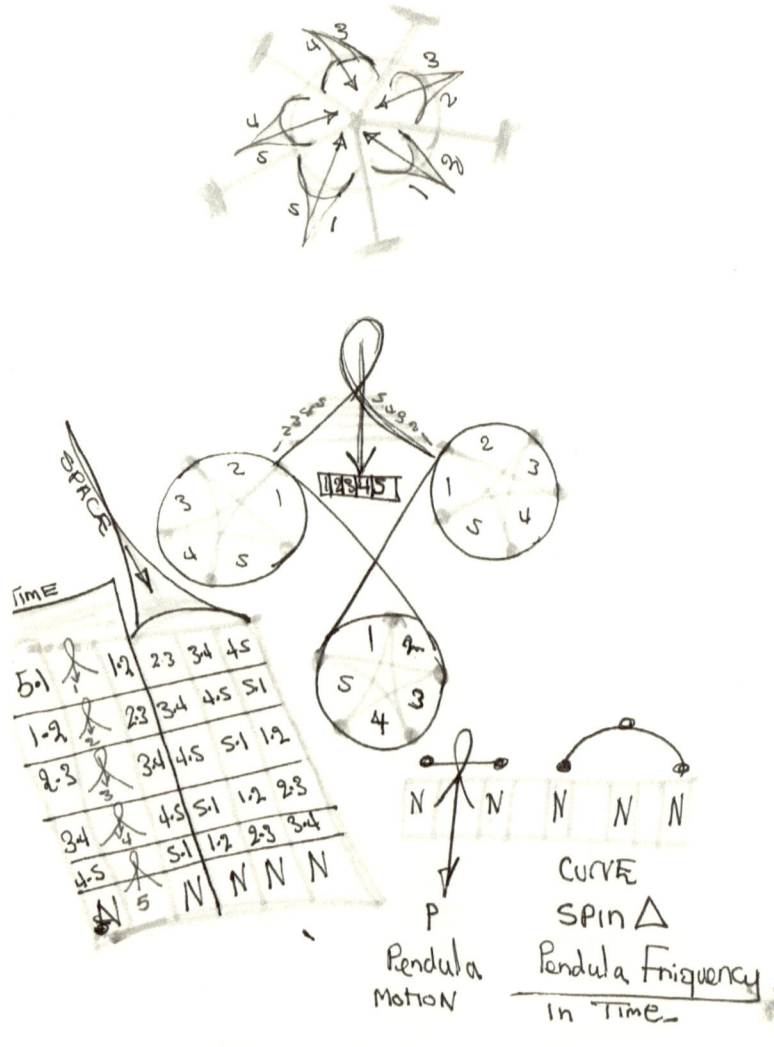

Mr. Farek J Sadon
Organun Company Director

Education & qualifications
- BSc Business and Economic Management
- HND Computer Studies
- DIPLOMA in the Law of International Displacement International Institute of Humanitarian Law SANREMO

UNHCR Certificates
- Refugee Registration
- Refugee Status Determination
- Refugee Reintegration
- Immigration
- Humanitarian Cooperation

Skills and achievements
- US 404th Affairs Battalion Certificate of Appreciation for assisting IDPs and refugees
- Award of Korean Mandala medal for humanitarian coordination with South Korean Military Civil Affairs
- Languages; Fluent in English, Arabic, Farsi(Iranian) & Kurdish(Mother tongue)
- IT skills; Hard, Soft Engineer
- Customer service & Administration skills
- International Political Arena Director and Skills
- Active Member in Human right watch Middle East, and International arena

Technical Experience

1989- 2000
Computer Engineer Worked with or in coordination with Apple London Centres
Science and Information Technology
Contributed in Design and translation of Arabic Interface Apple Mac Desktop Publishing
That include: First Arabic News Paper Manual to Digital Design
Contributed: in Production Design of all most all Famous Arabic Newspapers
    a. Saudi Research & Marketing Computer Graphic Designer for Newspapers & Magazines Al-sharq Al-Awsat plus Magazines
    b. Al-Hayat News Paper Al Qabas Kuwait and tens of other Arabic And Farsi, Turkish, Publication and Radio TVs Stations
    c. Contributed in Producing a computer Dictionary the News Net devise which is employed by British and other international new papers and publication Centres
    d. Made researches and efforts to find a standard Global method and a text book as computer technology tuition system continue working on that
    e. Made researches in RBG, CRT, screen Resolution and communication speed, Laser, sonic system and other means of data compression and Internet system

2000 to date;
Involved in Iraqi and International Politics Since the age of 16
As a diaspora I was given the opportunity to work for the ministry of humanitarian aid & cooperation Erbil Iraq the ministry, Established under UN Mandate

2002-2008
United Nations Office for Project Services - Iraq
Liaison Officer

Ministry of Humanitarian & Cooperation - Erbil- Iraq
Liaison with UN on Oil for food Program

Director General of IDP and Refugees Issues
Director of International Relations for humanitarian Issues
Counsellor and fund raiser for the Kurdish Academy and Artists UN in Erbil

Mobilised international assistant for IDP reintegration and rehousing.

Reported any Human right violations committed by tribal and sectarian militia forces to the human right watch and other international Humanitarian organisations

Organised Campaigns against corruption and luting of Public Funds committed by tribal Militia forces.

Contributed in the organization of Transformation Movement in Kurdistan GORAN

Other Duties included:
- Coordination with UNHCR, UNOPS, NGOs, US Civil Affairs and Humanitarian Organisations
- organising IDP groups, visiting their locations, holding meetings, writing reports, translation and interpretation of documents and Information

www.organun.co.uk
+447466674926

www.ingramcontent.com/pod-product-compliance
Lightning Source LLC
Chambersburg PA
CBHW030838180526
45163CB00004B/1368